The Politics of Mapping

SCIENCES

Geography and Demography, Field Director – Denise Pumain

Political Geography, Subject Head – Frédéric Giraut

The Politics of Mapping

Coordinated by

Bernard Debarbieux

Irène Hirt

iSTE

WILEY

First published 2022 in Great Britain and the United States by ISTE Ltd and John Wiley & Sons, Inc.

ISTE Ltd
27-37 St George's Road
London SW19 4EU
UK

www.iste.co.uk

John Wiley & Sons, Inc.
111 River Street
Hoboken, NJ 07030
USA

www.wiley.com

Library of Congress Control Number: 2021951707

British Library Cataloguing-in-Publication Data
A CIP record for this book is available from the British Library
ISBN 978-1-78945-067-5

ERC code:
SH2 Institutions, Values, Environment and Space
SH2_1 Political systems, governance
SH2_11 Human, economic and social geography
SH2_12 GIS, spatial analysis; big data in political, geographical and legal studies

Contents

Introduction

Bernard DEBARBIEUX and Irène HIRT
University of Geneva, Switzerland

I.1. Theoretical and conceptual debates

The focus on the relationship between mapping[1], on the one hand, and politics or power, on the other hand, is certainly not new, or even recent. For centuries, particularly in the Christian, Chinese and Arab-Muslim worlds, the people who commissioned maps (princes, ministers or heads of state) and the cartographers clearly believed that maps must, first and foremost, serve the objectives of knowledge and the control of territories by sovereigns, armies and state administrations.

In the Western world, the proximity between maps and political power became less visible around the 18th century, when the discourse among cartographers began to focus on the techniques of production, the accuracy of the representation and the scientific value of the knowledge produced. This shift became even more pronounced in the 20th century when the scientific analysis of the map corpus and reflection on

1 The difference of meaning between mapping and cartography is not stabilized even among specialists of the field. In this introduction, in accordance with the title of this book, we focus on mapping broadly defined as the way of thinking, seeing and acting by using maps; comparatively, cartography points more specifically at "the art, science and technology of map-making" which emerged as a scientific discipline in the late 19th century (Kent and Vujakovic, 2018, p. 1). However, some well-known phrases in the field have used the word "cartography" in a different way, such as "critical cartography" or "history of cartography". In this case, we have kept the word cartography to respect the common use of authors and practitioners. For the sake of convenience, we will also use words derived from "cartography" (cartographer, cartographic production, etc.) interchangeably, i.e. both when referring to "cartography" and "mapping".

map production itself became essentially guided by a belief in the progress of techniques and knowledge. This led to the triumph of verism in the representation of the Earth's surface, accuracy in the representation of data and optimization in the communication of knowledge. This threefold objective was achieved through the increasing sophistication of modern societies, techniques and policies. A few key figures in cartographic thinking in the second half of the 20th century embodied this ambition. Among these figures, Bertin (1983 [1967]) reflected on the visual and semiological effectiveness of maps. However, from the 18th to the 20th centuries, this discourse on the efficiency of maps was often voiced by authors working directly or indirectly for political authorities, sometimes even within their institutions. The added value maps held for the exercise of power tended to escape critical reflection, since it became consubstantial with cartographic knowledge, on the one hand, and with government know-how, on the other hand.

I.1.1. *The "Harleyian turning point"*

The emergence of an alternative research program with this type of reflection as its objective can be traced back to the 1970s and 1980s. Referred to as "critical cartography", it is largely in line with the French, Italian, Portuguese and Spanish approaches to the "social and political uses of maps". This program proposes a critical interpretation of cartographic production in history, highlighting the socially constructed dimension of the map and its capacity to support strategies and relationships of power/knowledge as conceptualized by Michel Foucault. The break with the hitherto dominant approaches is remarkable. The main innovation of the 1970s and 1980s involved breaking away from the modernist and evolutionist narrative to question that which had been largely left out of the field of reflection, namely the political, but also social and contextual dimensions of cartographic commissioning and production. The historian of cartography Brian Harley emerged as a leading figure in this breakthrough (Blakemore and Harley 1980; Harley 1988a, pp. 277–312, 1988b, 2001). The publication of his article "Deconstructing the Map" (1989) had a major influence on thought about the "power of maps". His work has become a classic despite its theoretical and methodological shortcomings, which are noted by even Harley's greatest advocates, as Reuben Rose-Redwood (2015) pointed out in the special issue of the journal *Cartographica* published in 2015: "Deconstructing the Map: 25 years on".

This turning point is all the more significant because Harley was followed by many authors who produced a significant body of empirical or theoretical analysis. Even today, his proposals continue to inspire researchers. Indeed, referring to the idea of "critical cartography" quickly became a standard, particularly in the English-speaking academic world, at a time when geographical analyses in terms of power relations and domination were also multiplying. From that moment on, the political dimension of critical cartography has tended to be emphasized (Crampton and Krygier 2006, p. 11). This has resulted in considerable and exciting academic output. Initially, this focused primarily on the role of mapping both in the history of modern state-building in Europe (Buisseret 1992; Kain and Baigent 1992) and colonial empires (Kain and Baigent 1992; Edney 1997; Blais 2014) – leading, in the process, to the negation of Indigenous peoples, especially in the Americas (Harley 1990, 1992) – as well as in the process of decolonization (Akerman 2017). However, it has also looked at the diffusion of the modern state model outside European colonies (Winichakul 1994).

On the strength of these case studies and the constant enrichment of the field that has been made possible, critical cartography acquired significant visibility in the last two decades of the 20th century, thanks to a number of exhibitions[2] and critical works, including Denis Wood's *The Power of Maps* (1992), Mark Monmonier's *How to Lie with Maps* (1991), and the ambitious collection of volumes, *The History of Cartography*, launched in 1987 by Harley and David Woodward (1987), to which we will return later. Critical cartography has also become globalized through the translation of some of Harley's texts[3] and the publication of analyses made by authors whose approaches were similar to his[4] but who write in languages other than English (e.g. Jacob 1992; Gugerli and Speich 2002; Valerio 2007). This visibility has enabled maps and mapping to occupy an important place in the work of major authors on the issues of nationalism (Anderson 2006 [1983]) and the state (Scott 1998), despite them being outside the field of critical cartography itself.

However, this account of the critical and political turn taken by cartographic thinking, while bringing to light a very important moment in the history of mapping, which was stabilized by being endlessly revisited by academic production, is not fully satisfactory, nor is it complete. There are several reasons for this.

I.1.2. *Pluralism and taking perspectives further*

First of all, this narrative pays little attention to the reflections and work that preceded the "Harleyian turning point". Throughout the 20th century, and well before the 1970s, geographers emphasized the use of maps for political purposes.

2 Notably the exhibition *The Power of Maps*, created in 1993 at the Cooper-Hewitt, National Museum of Design, New York, and later hosted at the Smithsonian Institution, Washington.

3 In French, the translation and editing of Harley's few texts were done by Peter Gould and Antoine Bailly (Harley 1995).

4 Monmonier's most popular work (1991) was translated into many languages in the years 1990–2000.

Isaac Bowman, for example, director of the *American Geographical Society* and *Chief Territorial Specialist*, appointed by Woodrow Wilson to advise the American delegation to the Paris Conference in 1919, was astonished by the quantity of contradictory linguistic and ethnic maps placed on the negotiating table to decide the European borders in the aftermath of the First World War, each illustrating the claims of the states concerned (House and Seymour 1921; Smith 1984). In the aftermath of the Second World War, Hans Speier and a few others analyzed Nazi propaganda maps, showing the fundamentally rhetorical and persuasive nature of maps (Speier 1941)[5]. Finally, authors relying on Marxist theories, such as Yves Lacoste with the topographic map (Lacoste 1976), have endeavored to show how mapping has served the objectives of military, social or imperialist domination. In comparison with the production of previous decades, the "Harleyian turning point" can hardly be characterized by the mere (re-) discovery of the political nature of maps.

Besides, this "turning point" is not characterized by a single problematization of the coupling between maps and politics nor by a stable body of theoretical references. It is true that Harley favored a critical interpretation of the map very early on; in doing so, he was concerned with placing it in its social context of production. However, at first, this interpretation was also mainly fed by Marxist theories and questioning (as in Harley 1988a, pp. 277–312). It was only gradually, under the influence of very different postmodern authors – Roland Barthes, Jacques Derrida and Michel Foucault (who we are still surprised to see grouped under the single banner of "French theory" in the English-speaking world) – that Harley called for a research program aimed at "deconstructing" (Harley 1989) the map, "denaturalizing" it in the opposite direction to what the modernist project had sought to do, and recognizing within it intentions that were indistinguishable from the productions. In this respect, Harley's work is not isolated. It is nourished by numerous new reflections on maps and mapping, as shown by the reference to the catalog of the exhibition "*Cartes et figures de la terre*", organized in Paris in 1980 (Centre Georges Pompidou 1980), in his first theoretical text on the subject (Blakemore and Harley 1980). Moreover, this work was undertaken in parallel with others who drew, in part, on the same theoretical references: this is the case of Claude Raffestin, who was undoubtedly the first European geographer to take advantage of the Foucauldian method to analyze geographic and cartographic discourse, for example, when he wrote: "The zenithal gaze is dominant by nature and is inscribed in the desire to see everything in order to know everything and to be able to do everything" (Raffestin 1988, p. 136 author's translation; see also Raffestin 1979).

Drawing on Foucault's early work, Harley emphasized the disciplinary and authoritarian potential of state mapping. Many of those who have extended his work,

5 Speier's text has been commented on since the turn of the century, notably by Pickles (2004, pp. 44–45) and Herb (1997).

however, have favored Foucault's second generation of work, particularly that devoted to the notion of governmentality. In doing so, attention shifted from topographic and political maps to thematic maps, which better reflected the need for knowledge and the control of populations and were even more easily manipulated than topographic maps (Raffestin 1988, p. 137). With this view, John Pickles wrote in 2004 that "mapping as a power-knowledge also functions as productive power to constitute objects, identities and practices that are part of (and constitutive of) our world" (Pickles 2004, p. 114). Mapping is therefore of interest as much for its contribution to the territorialization of state power as for its participation in the adoption and dissemination of hegemonic concepts of territoriality. Such analyses lead several authors to think of the map as a condition for the territoriality of the modern state (Winichakul 1994; Kitchin and Dodge 2007; Debarbieux 2019), and for that of modern societies as a whole. Pickles himself puts it this way, "Maps and mappings precede the territory they 'represent'" (Pickles 2004, p. 5).

This kind of thinking has led Joe Painter to speak of "cartographic anxiety" in order to refer to the need of states (but also any other political entity) to have a territory whose very nature can be shaped by maps: "I want to suggest that dominant understandings of concepts such as 'territory' and 'region' have been structured in important ways by cartographic reason, and therefore that related practices of territorialization and regionalism are generated, in part, by cartographic anxiety" (Painter 2008, p. 346). At this point, the interpretation of the map as an instrument of power has seen its epistemological character coupled with an ontological one: it participates in an ontogenesis of the state and the corresponding society. In this book, Franco Farinelli's chapter on the role of mapping in the spatial production of modern societies (Chapter 2) and Bernard Debarbieux's chapter on the cartographic production of modern states (Chapter 5) attempt to explain and illustrate this perspective.

However, this thesis only holds if it is supported by a detailed analysis of the system of signs of which a map is constituted, or even by a theory of political representation. That the map requires an organized system of signs is a given, even before the invention of the map legend, at the end of the 17th century (Harvey 1993). The constant improvement of this system was the main preoccupation of those who, from Bertin to MacEachren, strove to optimize the map's capacity to communicate information. However, the analysis only takes on a critical and political dimension when one looks at the selection and prioritization of mapped objects and the political effects of these choices. Harley himself distinguished between the "external power" of maps and mapping – the power exercised *over* them through political commissioning, or exercised *with* them, through their political, administrative, religious or legal use – and the "internal power" of maps and mapping (in the same way as one might speak of the power of words), namely, their performative character and their political effects, engendered, among other things, by the choices made in representation (Harley 1989, pp. 12–13).

The analysis of the maps produced following the voyage of Christopher Columbus, accompanying the colonization of the American continent by Europeans, is emblematic of the identification of this "internal power", including the invisibilization of the Indigenous populations and the concern taken to visually offer a space to be conquered (Harley 1992). It is to this "internal power", mainly linked with the choice and arrangement of signs, that the often-used expression "power of maps" refers (Wood 1992). In fact, it was long before 1992 and the peak of critical cartography, represented by the highly controversial celebration of the 500th anniversary of the "discovery" of America, that some people were busy subjecting the map to the prism of theories of representation. Already, in the 1970s, Louis Marin's detailed analysis of the components and general layout of Jacques Gomboust's 1652 map of Paris, including the text boxes, was a true statement of method. For Marin, this portrait of Paris is an "effect machine" among other effects, particularly effective in the simultaneous institution of what it designates (the city) and of the one who commands it (the King of France) (Marin 1973, 1980, pp. 47–54). Around the same time, Raffestin used the work of another representation theorist, Hillary Putnam – who claimed that "objects do not exist independently of conceptual frameworks" (Putnam 1984, p. 64, author's translation) – to remind us that the cartographer does not describe the real as it is, but constructs a system of objects that reflects an intentionality (Raffestin 1988).

Beyond English-speaking geographers and cartographers, it is those on the European continental mainland who seem to have taken the analysis of the "internal power" of the map the furthest. We can cite the works of Ollson (2007) and Farinelli (2009, see also Chapter 2 of this book) on "cartographic reason", in particular, the recourse of modern cartography to Euclidean geometry and its social and political effects. In this same vein, in the work of Casti (1993, pp. 79–101, 2005), and the research group she has built around her at the University of Bergamo, Italy, the analysis draws more specifically on critical semiology. At the same time, Jacques Lévy and his team have, since the 1990s, been engaged in a symmetrical exercise, this time an epistemological one: by postulating the existence of a "cartographic turning point", Lévy has been proposing avenues of map analysis regarding the political dimension of societies, while at the same time distancing himself from Euclidean metrics (see Chapter 1).

I.1.3. *Toward process- and practice-centered approaches*

At the same time, other theoretical and conceptual viewpoints have come to nourish a political analysis of the map. Many authors have pointed out that the reduction of the analysis of mapping to the question of representation has its own limits; they have advocated an analysis that takes into account the processes and practices of cartographic production (*mapping*) as much as the *map* as a product or artifact (Del Casino and Hanna 2006; Kitchin and Dodge 2007). Kitchin and Dodge (2007) have emphasized the need to question the "ontological (in)security" of maps, as they emerge through practices of territorialization. Pickles has sketched out the features of a pragmatics of the map by asking how it functions in practice. Drawing on the work of Michael Curry, he suggests that the map is "not a representation of the world, but an inscription that does (or sometimes does not do) work in the world" (Pickles 2004, p. 67). This hypothesis is reminiscent of the proposals of the sociology of science and technology, which, whenever concerned with the map, emphasizes the cognitive and political advantages that the map provides through its materiality, and through the circulation and accumulation of knowledge that it makes possible as an "immutable mobile", i.e. a stable object which travels, carrying along its own content and power of influence (Latour 1987, 2005; Turnbull 2000; November *et al.* 2010).

In fact, these "more-than-representational" or "processual" approaches had already been raised 20 years earlier by Robert Rundstrom, who had pointed out that textual or representation-centered approaches could hardly be applied to societies with an oral tradition (1991). This was followed by the famous book series *The History of Cartography* – in particular the introduction to Volume 2.3 on African, American, Arctic, Australian and Pacific societies (Woodward and Lewis 1998, pp. 1–10), which constitutes a reference for several chapters of this book – that reaffirmed this necessity by relying on the anthropological plurality of the act of mapping. The political dimension of this development is twofold: on the one hand, in epistemological terms, it put an end to what Harley himself had called "scientific chauvinism" (Harley 1987, p. 4) in cartography; in other words, it challenged the Western hegemony over the definition of what a map and what cartographic knowledge is. On the other hand, this "more-than-representational" interpretation deepened the analysis of the expressions of power; this is no longer limited to the study of the map's commissioners' intentions or to their social and political effects; it also deals with the relationships of power as they unfold in mapping processes and practices.

Such a conceptual expansion allows us to fully grasp the challenges of the digital and online turning point of mapping, in particular, the non-representational dimensions of digital geographic information, as well as its circulation and dissemination. "In fact, maps are no longer maps," says an article in the French newspaper *Le Monde*. "They are databases, i.e. sets of points, objects, names... which, on demand, are transformed into a particular map, with a different appearance according to needs" (Larousserie 2013, author's translation). Yet, while this type of data is multiplying (big geodata), becoming ubiquitous in people's daily lives (Global Positioning System (GPS) in cars, geolocation applications on phones, consultation of websites, etc.), it is becoming increasingly difficult to analyze the power relations that structure these new phenomena and their social and political implications. This is why, according to Pierre Gautreau and Mathieu Noucher (Chapter 3), understanding these "shifting" objects implies reinventing the concepts and methods of critical cartography. The question of intentionality becomes particularly complex to decipher since the distinction between amateur and expert, or between producer and user of a map (or of geographic information) often proves to be blurred (we then talk about "produsers") (Noucher 2017; Gautreau and Noucher in this book). Finally, the contributions of pragmatics and semiology, attached to the detailed observation of the signs and micro-actions involved in cartographic production, have also consolidated the empirical validation of the critique, which has been said to be lacking in many previous works (Palsky 2003)[6].

Finally, the range of theories and concepts that have been mobilized to think about and operationalize a critical cartography of state and colonial production has expanded considerably over time. The Marxist critique of the 1970s, which was in some ways the inaugural one, gave way to more detailed analyses of the relationships of power and domination as expressed in cartographic production, while also turning to questions of identity and culture (identity politics), under the influence of cultural studies and post-colonial and decolonial criticism. The problem of gender, on the other hand, has long constituted a silence in the approaches that are part of the "Harleyian tradition". Indeed, while Harley was particularly sensitive to the intentional and unintentional silences of maps (Harley 1988b), he himself remained silent on the role of women in cartographic production and on one of the most structuring categories of social relations: gender. According to Nikolas Huffman, this blind spot is due to conceptions of society and space that have long since naturalized the masculinist episteme of science, and thus also of mapping (Huffman 1995, pp. 1622–1630).

This book, through its above-mentioned chapters, aims to give an account of the variety of political analyses of mapping (and geographic information) and their

6 Palsky points to certain commentaries on the "Cassini map" which argue that it was intended to constitute a body of knowledge capable of reinforcing the centralization of France in the 18th century and its military power, whereas specialists of this production have shown how it was the result of contingent decisions, always thwarted in its realization (Palsky 2003, p. 212).

theoretical or methodological references. It goes beyond the perimeter of critical cartography by not limiting its scope to relationships of power and domination. While it takes into account the latter (politics), it does not ignore the institutional issues (polity) that effect the production of maps, nor the public policies that give rise to many maps, certainly in order to justify and design them, but also to distribute them (see Chapter 9)[7].

I.2. Political practices and uses of maps

The second objective of this collective work is to point out a series of social and technical transformations that force us to move away from the implicit equivalence between mapping, power and the state. In addition to the fact that the political analysis of societies has, on the whole, become detached from a vision centered on the state, fundamental social transformations have taken place over the last few decades in the production and use of maps and geographical information.

I.2.1. *New modalities, new actors: the questioning of cartographic state sovereignty*

The second half of the 20th century saw a major change in this field: the diversification of actors gaining access to the production of maps, which had, since at least the 19th century, been monopolized by state and military institutions. The launch of the first satellites (see Sputnik 1, sent into space by the Soviets in 1957) marked the first challenge to the cartographic sovereignty of states: indeed, outer space is marked by a legal vacuum (it refers to space beyond airspace, the latter being included in the sovereignty of states). Although the producers of satellite images were initially guided by military uses, from the 1970s onwards, remote sensing expanded its civil uses (Landsat, Spot images, etc.). However, data production has remained permanently dominated by the United States and a few other hegemonic actors (Desbois 2015). Hy Dao (see Chapter 7) shows that international governance arising after the Second World War (United Nations (UN) agencies in particular) has also adopted the production of a certain type of data, supposedly to help solve major global issues, including environmental problems.

7 In French, the word "*politique*" encompasses the following three concepts of the English language: 1) *Polity* or "*le politique*": the institutional dimension of politics; the institutions of the social world; 2) *Politics* or "*la politique*": all the power relations of the social world, and not only those inherent in the institutions; 3) *Policies*: "*politiques publiques*" the set of laws and instruments which orientates the action of governments.

The second challenge to the state's quasi-monopoly in cartographic and geographic information production occurred at the turn of the 21st century, with the digital revolution and the widespread use of the Internet and geospatial technologies. Global players such as Google benefited from the transfer of digital geographic information technologies from the military to the civilian domain (Cloud 2002; Desbois 2015). Not only have they challenged the monopoly of states in the production and dissemination of geographic information, especially on the Internet, but they have also entered the daily lives of individuals. These private companies now possess significant means of cartographic production – or, more generally, of georeferenced information – and contribute massively to the "deluge" of data and images (the famous big data).

The third crack in the edifice of state cartographic sovereignty has come "from below". As Jeremy Crampton and John Krygier point out: "In the last few years cartography has been slipping from the control of the powerful elites that have exercised dominance over it for several hundred years" (Crampton and Krygier 2006, p. 12). Grassroots social movements or civil society actors with "local" anchors ("inhabitants", "citizens", members of cultural "minorities" or "Indigenous peoples", etc.) have been increasingly producing their own maps (Debarbieux and Lardon 2003). The local framing of their actions has not prevented them from defending causes that are sometimes transnational or even global (as in the case of the fight against climate change). This shift may have taken place both because of easier and cheaper access to certain mapping or geospatial techniques and because of their growing involvement in development or territorial planning projects. These actors have seen cartographic language as a means of defending the quality of their space and living environment, of claiming rights, of promoting political causes or seeking sociospatial justice; or, in a less radical spirit, as a means of dialogue with the authorities in order to make their visions and perceptions of space known. These processes are part of both resistance and openness attitudes, as Greg Brown and Hunter Glanz show when describing participatory GIS practices in the context of territorial planning projects in the state of California: the mapping of their spatial preferences by inhabitants highlights both negative attitudes of rejection (nimby or Not-In-My-Back-Yard) and positive attitudes of support for these projects (yimby or Yes-In-My-Back-Yard), depending on the distance between their home and the project (Brown and Glanz 2018). Another characteristic of these actors "from below" is that they have massively taken on new digital technologies and the geoweb, both software and programs produced and controlled by the GAFAs, as well as free software. However, the political effects of this investment have not always been what one might expect. For example, the OpenStreetMap (OSM) free maps, produced by thousands of volunteers, have now become a reference for political authorities, especially municipalities in France, whose data from official maps are generally not as up-to-date as those from the OSM.

I.2.2. *From domination to resistance strategies*

The "power of maps" has thus shifted from the sole issue of domination and hegemony to the more general question of power relationships and strategies of resistance, supported by maps or geographic information. Here again, the limitations of the framework of thought proposed by Harley, whose approach was essentially state-centric, neglecting alternative expressions of mapping opposing hegemonic forms of power, may be noted (Crampton 2001, p. 236; Dodge and Perkins 2015; Gautreau and Noucher, Chapter 3 of this book). Indeed, in 1988, Harley (1988a) wrote: "The social history of maps, unlike that of literature, art, or music, appears to have few genuinely popular, alternative, or subversive modes of expression. Maps are pre-eminently a language of power, not of protest" (p. 301). But perhaps Harley died too soon (1991) to observe the proliferation of protest mapping practices, which, although existing for several decades, only really took off in the 1990s. As a historian, moreover, observing social reality "in the making" was not his primary object of study.

Some of the first expressions of "cartographic resistance" came from intellectuals or artists who wanted to challenge Eurocentric worldviews. The famous controversy launched by the German historian Arno Peters is emblematic of this resistance. Very critical of the so-called Mercator projection, which over-represents the rich countries of the "North" by exaggerating their surface, in 1973 Peters proposed a projection that respects the correct size of the continents relative to each other (but not their shapes)[8]. Giving more weight to the countries of the "South" in the representation of the planet, this projection was used worldwide, and was quickly adopted by Third Worldists fighting against inequalities between countries, by church and development aid organizations, by UN agencies and schools. The cover page of the Report of the Independent Commission on International Development, chaired by Willie Brandt (Figure I.1), is an example of the institutional use of the Peters projection. A North-South dividing line has been added to the map to emphasize the surface importance of the "South". As Jean-François Staszak, Juliet Fall, and Frédéric Giraut point out, however, "the designations north and south are partly metaphorical since much of Oceania is in the north while the Manchu confines are in the south" (Staszak *et al.* 2017, p. 163, author's translation).

8 This projection is also known as the Gall-Peters projection, after the Scottish minister James Gall who invented it in the 19th century.

Figure I.1. *Cover page of the report, "North-South. A Programme for Survival" (source: Brandt and Independent Commission on International Development Issues (1980)). For a color version of this figure, see www.iste.co.uk/debarbieux/mapping.zip*

As Timothy Barney has pointed out, in the tumultuous decolonization and Cold War context of the 1970s, this projection was more a rhetorical event than a technical innovation, presenting the map as a powerful medium for change, rather than a neutral representation of space (Barney 2014, p. 104). Barney has suggested, however, that despite the radical nature of its project, this projection nonetheless also contributed to a "recolonization" of the world, supporting the logics of Western liberal development in the so-called "Third World" (Barney 2014).

Long before Peters, the field of art had appropriated the map as an iconographic medium for an openly ideological discourse of decolonization of knowledge and representations of the world. *América Invertida,* drawn in 1943 by Joaquín Torres García (1874–1949), a Uruguayan and Catalan artist who created the "universal constructivism" movement, presents South America in reverse of the conventions orienting maps toward the North (Figure I.2). In addition to the Earth's equator marking the separation between the northern and southern hemispheres, Torres García added the latitude and longitude of the city of Montevideo (marked with a cross) where he was born and lived for part of his life, making the drawing a deliberately situated representation. Seeking to inaugurate an artistic movement rooted in South America and which would come as a counterpoint to European influences, the artist denounced the domination of the countries of the North over those of the South: "[...] There should be no North for us, except in opposition to our South. […]. This is a necessary rectification; so that now we know where we are[9]" (Rommens 2018, p. 36).

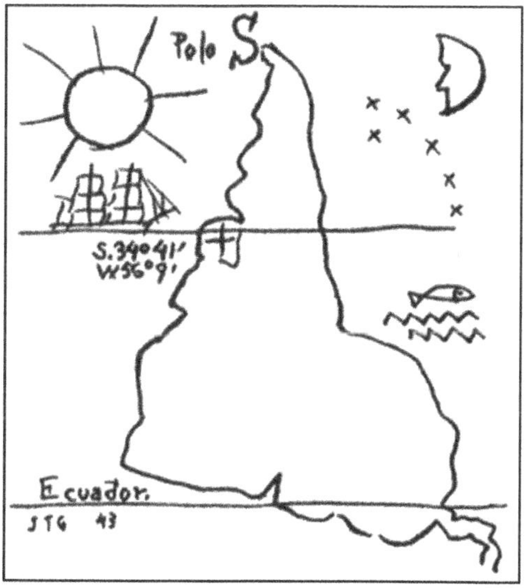

Figure I.2. *"América invertida" (Inverted America), 1943 (Museo Juan Manuel Blanes, Montevideo). Drawing (ink on paper, 19.5 x 15.5 cm) (source: Joaquín Torres García)*

9 We have not been able to find the original text published in Spanish in García Torres' manifesto *Estructura* (1936), where an early version of the map appears.

While Torres García's cultural and artistic intentions and influences are debated[10], the effects of this image and the extent of its circulation are interesting in the context of a critical cartographic analysis that deciphers the power of maps to forge worldviews. The drawing is now known more for its political impacts than for its artistic or aesthetic qualities; it still continues to forge an imagining of Latin America and its identity today, asserting the idea of Latin American cultural and political independence and challenging the hegemonic "center-periphery" or "North/South" model of the world (Rommens 2018). In addition to the fact that the drawing is mobilized in Latin American academic circles that develop decolonial or postcolonial approaches (conference posters or seminars), it is very popular on the continent, reproduced on T-shirts, tourist souvenirs of Uruguay, posters and postcards, including on tattooed bodies, as a quick search on Google images shows (Figure I.3). Each of these mobilizations gives rise to new meanings and uses (Monteiro Sales 2016).

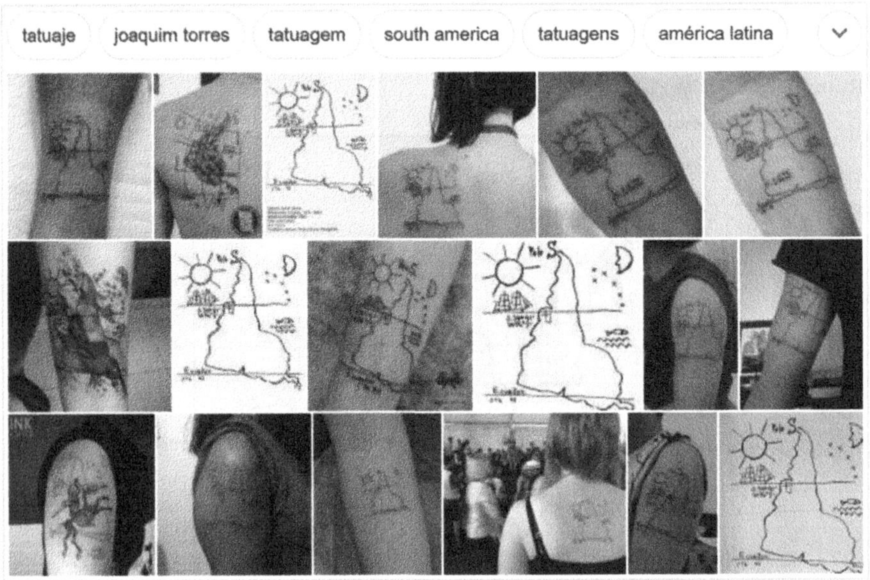

Figure I.3. *Screenshot taken from Google images with the keywords "América invertida tattoo" on February 4, 2021. For a color version of this figure, see www.iste.co.uk/debarbieux/mapping.zip*

10 For an in-depth interpretation of this artistic map and the artist's intentions (between a desire for decolonization and unassumed cultural mixing), see Rommens (2018).

The same spirit can be found in the comic book writer Joaquín Salvador Lavado's work, known as "QUINO": his main character Mafalda, who became a cult in Argentina, and whose stories were published between 1963 and 1974, has a very critical and pessimistic view of society and the state of the world. In an excerpt from the comic strip, the little girl looks at a globe with astonishment, discovering that the inhabitants of the Southern Hemisphere live upside down. Drawing parallels between cartographic representations and the epistemological domination of the South by the North, she concludes that "Since we are living upside-down, the ideas fall off of us" (Lavado 2018 [1964], pp. 99–100).

In a way, these manipulations of the constitutive principles of maps (projection, orientation and graphic language) for ideological purposes or cultural decentering are an extension of Harley's own reflections in *Deconstructing the Map* on the "rule of ethnocentricity" of world maps (Harley 1989, p. 6). These diversions were innovative and powerfully evocative during the Cold War period, when questioning a Western-centric view of the world was not self-evident. Today they have become common practice. However, their pedagogical effectiveness continues to be proven today, leading students to question the conventionality of their worldview.

As for marginalized social groups or social movements seeking to make their voices heard on the political scene (be it national, regional or international), their mobilization of "protest maps" (Wood *et al.* 2010) is aimed, above all, at influencing public policies on land use planning, land tenure or the management of natural resources, which are deemed to be unjust. This specific use of maps led Nancy Peluso (1995) to coin the neologism "counter-mapping". As Irène Hirt explains in Chapter 7, Indigenous peoples were among the first social groups to make use of state language, techniques and modes of representation, as early as the 1950s and 1960s, with all the ambiguities that this implies for groups claiming to have a world view distinct from that of the state. Since then, other social and political actors have resorted to this strategy (see (Kollektiv Orangotango 2018) for a range of such practices). Indeed, the term has been so successful that it is now also used to describe protest mapping practices of a more symbolic or rhetorical nature, such as those described above.

Easy access to the production of cartographic representations by individuals or organized groups in civil society also has the virtue of encouraging public debates around societal issues. Charles Booth's poverty maps of London in 1889 are a very early illustration of this (Topalov 2004). Olivier Clochard, in Chapter 10, illustrates the potential of maps to address contemporary migration issues. He describes how the Migreurop collective of researchers and activists produces its own maps in order to provide the public with alternatives to media and political representations of migration issues, to "oppose unjust migration policies", and to challenge the scientific standards of maps through creative means of expression (e.g. drawing).

I.2.3. *The issue of participation*

As Peluso has pointed out, this does not mean that mapping has gone from being a "science of princes" to a "science of the masses" (Peluso 1995, p. 387). Despite easier access to the production of maps and geographic information, it is still generally the preserve of individuals or social groups with good technical, economic, social and spatial capital. The digital divide is also a significant factor of inequality. Finally, these limits to the "democratization" of maps are, to a large extent, located in the issues of participation, i.e. in the ways in which the "public" is involved in the design of maps (Hirt and Roche 2013); these modalities being at the center of the relationship between maps and mapping, on the one hand, and power or politics, on the other hand. From the 1970s and 1980s onwards, these participatory mechanisms were instrumental in giving more people access to map production (see Chapter 4). However, several decades of experimentation and reflexive and critical analysis show that they are not self-evident.

From a critical and political perspective, Giacomo Rambaldi posed a fundamental question in this regard in 2005, one that is too often forgotten: "who owns the map legend?" (Rambaldi 2005). This is certainly the starting point for any participatory approach aimed at reclaiming the content of a map and what is represented, as Sylvie Lardon suggests in this book (Chapter 8). In this respect, recent experiments, such as the one conducted by Sarah Mekdjian and Anne-Laure Amilhat Szary, are particularly interesting. In collaboration with artists and a group of people of immigrant backgrounds in Grenoble, France, the two geographers retraced their migratory routes, emphasizing their lived dimension. To do this, the workshop participants developed a common legend using stickers and colors (expressing emotions, key events in the migration journey, etc.) that could transcend the languages spoken by the participants (Mekdjian and Amilhat Szary 2015). This project illustrates that the appropriation of the legend, in addition to its content, is also promoted by the way it is represented.

Moreover, it is the nature and quality of involvement that has been particularly debated in the literature on participatory mapping or GIS processes, regardless of their name: participatory mapping, participatory GIS, community-based participatory mapping, etc., both in the scientific field (in French, see the special journal issues edited by Noucher (2013) and Lardon and Noucher (2016)) and in the area of land and regional planning (see manuals of "good practices", such as (Corbett 2009) or (Gambier *et al.* 2016)). Participation in the field of mapping is usually measured by the degree of empowerment it fosters in terms of the transfer of skills from the expert cartographer to the "inhabitant", the latter's decision-making power in the production of the map, from the choices made upstream on what is mapped, to the way it is represented and communicated downstream. It also relates to the access and control of the information produced. These modalities are measured by tools that have extended Sherry Arnstein's (1969) famous "ladder of citizen participation" for assessing degrees of public involvement in decision-making to participatory mapping or GIS projects (Amelot 2013). However, in the face of the complexity and diversity of the forms of social order, participation is certainly still more of a goal to strive for than an actual reality, and is even sometimes an ideal imposed, once again, by the North on the South, as Xavier Amelot points out when he emphasizes the "participatory illusion" of mapping projects carried out in Madagascar (2013).

In the English-speaking world in the mid-1990s, a critique entitled "GIS and society" (Craig *et al.* 2002) also emerged, focused on the social implications of how people, space and environment are represented in GIS. This led to work on the more specific impacts of GIS on participation, power relationships, processes of inclusion or exclusion, and existing inequalities in access to digital technologies and data, and on the determinant character of social and political forces. It is within these debates that gender issues were first explicitly posed in critical cartography (Kwan 2002a, 2002b; McLafferty 2002; Schuurman and Pratt 2002; Pavlovskaya and St. Martin 2007; Elwood 2008).

The relationship of mapping with the more general issues of contemporary, interactive, collaborative or integrative democracy, instead of representative democracy, are highlighted by several of the contributions in this book. Sylvie Lardon (Chapter 8) presents a participatory prospective diagnosis protocol that she calls the "territory game" and that she has been implementing for the past 20 years in various contexts, through the mobilization of "choremes", the elementary structures of space invented by Roger Brunet in the 1980s. Federica Burini (Chapter 4) highlights the added value of information relating to the lived space of the "inhabitants" in order to know a territory through a semiological analysis of mapping. She includes in the notion of "participation" that which is, strictly speaking, not generally considered as such: i.e. the unconscious, or at least often involuntary, contributions of individuals through their smartphones and other connected objects, or by browsing the Internet. However, the possible effects of this geographic information are no less political, as it can be used by different actors for research, public policy making or even marketing purposes.

I.3. The structure of this book

Ultimately, 40 years after the publication of *Cartes et Figures de la Terre* (Centre Georges Pompidou 1980), and 30 years after the writings of Harley and other researchers, who furthered the critical approach to cartography, this book intends to contribute to the assessment of the work produced under this banner. It also aims to question the challenges posed by the technical, social and political changes in mapping, the forms and effects of which, although the subject of a growing number of analyses, have not yet given rise to large-scale syntheses. The preceding introductory elements were therefore intended as much as to sketch a genealogy of works questioning the relationship between mapping and "*politique*" (politics/polity/policies) as to underline the main issues that have run through them. Whether through the analysis of forms of representation or modes of production – authoritarian, participatory or contesting – they all enable us, to one degree or another, to see how mapping participates in "ordering the world" (Noucher *et al.* 2019), which concerns both the inhabited world and social and political worlds.

Apart from two of them which deal with the power effects of state cartography in a historical perspective (Farinelli and Debarbieux), the chapters that follow focus on the analysis of very current issues: the use of participatory mapping (Burini, Lardon), opening up the field of possibilities of mapping production to semiological and political innovations (Lévy), uses aiming at illustrating and defending the interests of Indigenous peoples (Hirt), the challenges posed by the explosion of data and digital mapping (Gautreau and Noucher), including in the work of international organizations (Dao), the use of alternative forms of representation to contribute to public debates (Clochard) and the changing modalities of state cartography applied to land use planning policies (Bailly).

However, these chapters take contrasting forms. The first group (Lévy, Farinelli, Gautreau and Noucher, Burini) is made up of essays: they are less concerned with offering a panorama of the academic production on the issues dealt with than with offering a personal reading. A second group (Debarbieux, Dao, Hirt) brings together texts that share a more encyclopedic concern; they attempt to identify the main orientations of the many works related with the chosen theme. A third group (Lardon, Bailly, Clochard) is made up of accounts of experiences: the authors are cartographers who are heavily involved in interactions with administrations, elected officials, groups of inhabitants and associations, and they show practical methods of mapping within particular political contexts: the past 20 years of cartography within the French administration of regional planning (Bailly), various territorial diagnosis and projects of local or regional scope (Lardon) and cartographic works aiming to illustrate migratory issues on modes capable of changing their representations (Clochard).

I.4. References

Akerman, J.R. (ed.) (2017). *Decolonizing the Map: Cartography from Colony to Nation*. The University of Chicago Press, Chicago.

Amelot, X. (2013). Cartographie participative pour le développement local et la gestion de l'environnement à Madagascar : empowerment, impérialisme numérique ou illusion participative ? *L'Information géographique*, 77(4), 47–67.

Anderson, B. (2006). *Imagined Communities: Reflections on the Origin and Spread of Nationalism*, revised edition. Verso, London, New York [Original published 1983].

Arnstein, S.R. (1969). A ladder of citizen participation. *Journal of the American Institute of Planners*, 35(4), 216–224.

Barney, T. (2014). The Peters projection and the latitude and longitude of recolonization. *Journal of International and Intercultural Communication*, 7(2), 103–126.

Bertin, J. (1983). *Semiology of Graphics: Diagrams, Networks, Maps*. University of Wisconsin Press, Madison [Original published in French 1967].

Bertin, J. (2013). *Sémiologie graphique : les diagrammes, les réseaux, les cartes*. Éditions de l'EHESS, Paris.

Blais, H. (2014). *Mirages de la carte : l'invention de l'Algérie coloniale, XIXe–XXe siècle*. Fayard, Paris.

Blakemore, M.J. and Harley, J.B. (1980). *Concepts in the History of Cartography: A Review and Perspective*. B.V. Gutsell, University of Toronto Press, Toronto.

Brandt, W. and Independent Commission on International Development Issues (1980). *North-South, A Programme for Survival: Report 11*. Pan Books, London.

Brown, G. and Glanz, H. (2018). Identifying potential NIMBY and YIMBY effects in general land use planning and zoning. *Applied Geography*, 99, 1–11.

Buisseret, D. (ed.) (1992). *Monarchs, Ministers, and Maps: The Emergence of Cartography as a Tool of Government in Early Modern Europe*. University of Chicago Press, Chicago.

Casti, E. (1993). Cartografia e politica territoriale nella Repubblica di Venezia (secoli XIV–XVIII). In *La cartografia italiana*, Casti, E. (ed.). Institut Cartogràfic de Catalunya, Barcelona.

Casti, E. (2005). Towards a theory of interpretation: Cartographic semiosis. *Cartographica: The International Journal for Geographic Information and Geovisualization*, 40(3), 1–16.

Centre Georges Pompidou (ed.) (1980). *Cartes et figures de la Terre : exposition présentée au Centre Georges Pompidou du 24 mai au 17 novembre 1980*. Centre Georges Pompidou, Centre de création industrielle, Paris.

Cloud, J. (2002). American cartographic transformations during the cold war. *Cartography and Geographic Information Science*, 29(3), 261–282.

Corbett, J. (2009). Cartographie participative et bonnes pratiques. Etude préparée pour le Fonds international de développement agricole (FIDA). *FIDA*, Rome [Online]. Available at: http://www.iapad.org/wp-content/uploads/2015/07/ifad_cartographie_participative_et_bonnes_ pratiques.pdf.

Craig, W.J., Harris, T.M., Weiner, D. (eds) (2002). *Community Participation and Geographic Information Systems*. Taylor & Francis, London, New York.

Crampton, J. (2001). Maps as social constructions: Power, communication and visualization. *Progress in Human Geography*, 25(2), 235–252.

Crampton, J. and Krygier, J. (2006). An introduction to critical cartography. *ACME: An International Journal for Critical Geographies*, 4(1), 11–33.

Debarbieux, B. (2019). *The Social Imaginary of Space*. Edward Elgar, Cheltenham.

Debarbieux, B. and Lardon, S. (eds) (2003). *Les figures du projet territorial*. Éditions de l'Aube and Datar, La Tour d'Aigues.

DelCasino, V. and Hanna, S. (2006). Beyond the "binaries": A methodological intervention for interrogating maps as representational practices. *ACME: An International Journal for Critical Geographies*, 4(1), 34–56.

Desbois, H. (2015). *Les mesures du territoire : aspects techniques, politiques et culturels des mutations de la carte topographique*. Presses de l'ENSSIB, Villeurbanne.

Dodge, M. and Perkins, C. (2015). Reflecting on J.B. Harley's influence and what he missed in "Deconstructing the Map". *Cartographica: The International Journal for Geographic Information and Geovisualization*, 50(1), 37–40.

Edney, M.H. (1997). *Mapping an Empire: The Geographical Construction of British India, 1765–1843*. University of Chicago Press, Chicago.

Elwood, S. (2008). Volunteered geographic information: Future research directions motivated by critical, participatory, and feminist GIS. *GeoJournal*, 72(3–4), 173–183.

Farinelli, F. (2009). *De la raison cartographique*. CTHS, Paris.

Gambier, S., Becks, B., Pedrick, C. (eds) (2016). Le pouvoir des cartes. Quand la 3D s'invite à la table des négociations [Online]. Available at: https://publications.cta.int/fr/publications/publication/1944/.

Gugerli, D. and Speich, D. (2002). *Topografien der Nation: Politik, kartografische Ordnung und Landschaft im 19. Jahrhundert*. Chronos, Zürich.

Harley, J.B. (1987). The map and the development of the history of cartography. In *The History of Cartography*, Harley, J.B. and Woodward, D. (eds). The University of Chicago Press, Chicago, London.

Harley, J.B. (1988a). Maps, knowledge, and power. In *The Iconography of Landscape: Essays on the Symbolic Representation, Design and Use of Past Environments*, Cosgrove, D. and Stephen, D. (eds). Cambridge University Press, Cambridge.

Harley, J.B. (1988b). Silences and secrecy: The hidden agenda of cartography in early modern Europe. *Imago Mundi*, 40, 57–76.

Harley, J.B. (1989). Deconstructing the map. *Cartographica: The International Journal for Geographic Information and Geovisualization*, 26(2), 1–23.

Harley, J.B. (1990). *Maps and the Columbian Encounter: An Interpretive Guide to the Travelling Exhibition*. University of Wisconsin, Milwaukee.

Harley, J.B. (1992). Rereading the maps of the Columbian encounter. *Annals of the Association of American Geographers*, 82(3), 522–536.

Harley, J.B. (2001). *The New Nature of Maps: Essays in the History of Cartography*. Laxton, P. (ed.). Johns Hopkins University Press, Baltimore.

Harley, J.B. and Woodward, D. (eds) (1987). *The History of Cartography*. The University of Chicago Press, Chicago, London.

Harvey, P.D.A. (1993). *Maps in Tudor England*. University of Chicago Press, Chicago.

Herb, G.H. (1997). *Under the Map of Germany: Nationalism and Propaganda, 1918–1945*. Routlege, London, New York.

Hirt, I. and Roche, S. (2013). Cartographie participative. In *Dictionnaire critique et interdisciplinaire de la participation*, Casillo, I., Barbier, R., Blondiaux, L., Chateauraynaud, F., Fourniau, J.-M., Lefebvre, R., Neveu, C., Salles, D. (eds). GIS Démocratie et Participation, Paris [Online]. Available at: http://www.dicopart.fr/fr/dico/cartographie-participative.

House, E.M. and Seymour, C. (eds) (1921). *What Really Happened at Paris; The Story of the Peace Conference, 1918–1919*. C. Scribner's Sons, New York [Online]. Available at: https://lccn.loc.gov/21008042.

Huffman, N. (1995). Silences and secrecy in the history of cartography: JB Harley, science and gender. In *Proceedings of the 17th International Cartographic Conference*. Institut Cartogràfic de Catalunya, Barcelona [Online]. Available at: https://icaci.org/files/documents/ICC_proceedings/ICC1995/.

Jacob, C. (2006). *The Sovereign Map: Theoretical Approaches in Cartography throughout History*. University of Chicago Press, Chicago [Original published in French 1992].

Kain, R.J.P. and Baigent, E. (1992). *The Cadastral Map in the Service of the State: A History of Property Mapping*. University of Chicago Press, Chicago.

Kitchin, R. and Dodge, M. (2007). Rethinking maps. *Progress in Human Geography*, 31(3), 331–344.

Kollektiv Orangotango (ed.) (2018). *This is Not an Atlas: A Global Collection of Counter-Cartographies*, 1st edition. Transcript Verlag, Bielefeld.

Kwan, M.-P. (2002a). Introduction: Feminist geography and GIS. *Gender, Place & Culture*, 9(3), 261–262.

Kwan, M.-P. (2002b). Is GIS for women? Reflections on the critical discourse in the 1990s. *Gender, Place & Culture*, 9(3), 271–279.

Lacoste, Y. (1976). *La géographie, ça sert, d'abord, à faire la guerre*. Éditions Maspero, Paris.

Lardon, S. and Noucher, M. (2016). Construire demain par les cartes : usages de l'information géographique en prospective territoriale participative. *Cahiers de géographie du Québec*, 60(170), 209.

Larousserie, D. (2013). Cartographie. La révolution numérique. *Le Monde*, 6 May 2013 [Online]. Available at: https://www.lemonde.fr/sciences/article/2013/05/06/ cartographie-la-revolution-numerique_3171889_1650684.html.

Latour, B. (1987). Les "vues" de l'esprit. *Réseaux*, 5(27), 79–96.

Latour, B. (2005). *La science en action : introduction à la sociologie des sciences*. La Découverte, Paris [Original published in French 1987].

Lavado, J. (2018). *Mafalda : intégrale*. Glénat, Grenoble [Original published 1964].

MacEachren, A.M. (1995). *How Maps Work: Representation, Visualization, and Design*. Guilford Press, New York.

Marin, L. (1973). *Utopiques : jeux d'espaces*. Éditions de Minuit, Paris.

Marin, L. (1980). Les voies de la carte. In *Cartes et figures de la Terre : exposition présentée au Centre Georges Pompidou du 24 mai au 17 novembre 1980*. Centre Georges Pompidou, Centre de création industrielle, Paris.

McLafferty, S.L. (2002). Mapping women's worlds: Knowledge, power and the bounds of GIS. *Gender, Place & Culture*, 9(3), 263–269.

Mekdjian, S. and Amilhat Szary, A.-L. (2015). Cartographies traverses, des espaces où l'on ne finit jamais d'arriver. *Visionscarto* [Online]. Available at: https://visionscarto.net/ cartographies-traverses.

Monmonier, M. (1991). *How To Lie With Maps*. The University of Chicago Press, Chicago.

Monteiro Sales, C. (2016). Cartografia, arte e visões de mundo na reprodução do "Mapa invertido da América do Sul". *Espaço e Cultura*, 157–178.

Noucher, M. (ed.) (2013). Cartographies participatives. *L'information géographique*, 77(4).

Noucher, M. (2017). *Les petites cartes du web. Approche critique des nouvelles fabriques cartographiques*. Éditions Rue d'Ulm and Presses de l'École normale supérieure, Paris.

Noucher, M., Hirt, I., Arnauld de Sartre, X. (2019). Ordering the world through maps and figures. Critical approaches to metrology [Online]. Available at: https://www.espacestemps.net/en/articles/ordering-the-world-through-maps-and-figures/.

November, V., Camacho-Hübner, E., Latour, B. (2010). Entering a risky territory: Space in the age of digital navigation. *Environment and Planning D: Society and Space*, 28(4), 581–599.

Olsson, G. (2007). *Abysmal: A Critique of Cartographic Reason*. University of Chicago Press, Chicago.

Painter, J. (2008). Cartographic anxiety and the search for regionality. *Environment and Planning A: Economy and Space*, 40(2), 342–361.

Palsky, G. (2003). L'esprit des cartes : approches historiques, sémiologiques et sociologiques en cartographie. HDR Thesis, Université de Paris XII-Val-de-Marne, Paris.

Pavlovskaya, M. and St. Martin, K. (2007). Feminism and geographic information systems: From a missing object to a mapping subject: Feminism and geographic information systems. *Geography Compass*, 1(3), 583–606.

Peluso, N.L. (1995). Whose woods are these? Counter-mapping forest territories in Kalimantan, Indonesia. *Antipode*, 27(4), 383–406.

Pickles, J. (2004). *A History of Spaces: Cartographic Reason, Mapping, and the Geo-coded World*. Routledge, London, New York.

Putnam, H. (1984). *Raison vérité et histoire* (translated by A. Gerschenfeld) Éditions de Minuit, Paris.

Raffestin, C. (1979). *Pour une géographie du pouvoir*. Librairies techniques, Paris.

Raffestin, C. (1988). Le rôle de la carte dans une société moderne. *Mensuration, photogrammétrie, génie rural*, 4, 135–139.

Rambaldi, G. (2005). Who owns the map legend? *URISA Journal*, 17(1), 5–13.

Robinson, A.H. (1952). *The Look of Maps: An Examination of Cartographic Design*. University of Wisconsin Press, Madison.

Rommens, A. (2018). Latin American abstraction: Upending Joaquín Torres-García's inverted map. *Mosaic: An Interdisciplinary Critical Journal*, 51(2), 35–58.

Rose-Redwood, R. (2015). Introduction: The limits to deconstructing the map. *Cartographica: The International Journal for Geographic Information and Geovisualization*, 50(1), 1–8.

Rundstrom, R. (1991). Mapping, postmodernism, indigenous people and the changing direction of North American cartography. *Cartographica*, 28(2), 1–12.

Schuurman, N. and Pratt, G. (2002). Care of the subject: Feminism and critiques of GIS. *Gender, Place & Culture*, 9(3), 291–299.

Scott, J.C. (1998). *Seeing Like a State: How Certain Schemes to Improve the Human Condition Have Failed.* Yale University Press, New Haven.

Smith, N. (1984). Political geographers of the past. *Political Geography Quarterly*, 3(1), 69–76.

Speier, H. (1941). Magic geography. *Social Research*, 8(3), 310–330.

Staszak, J.-F., Fall, J., Giraut, F. (2017). Les grands découpages du Monde. In *Frontières en tous genres : cloisonnement spatial et constructions identitaires*, Staszak, J.-F. (ed.). PUR, Rennes.

Topalov, C. (2004). Raconter ou compter ? L'enquête de Charles Booth sur l'East End de Londres (1886–1889). *Mil neuf cent. Revue d'histoire intellectuelle*, 22(1), 107–132.

Turnbull, D. (2000). *Masons, Tricksters and Cartographers: Comparative Studies in the Sociology of Scientific and Indigenous Knowledge.* Routledge, New York.

Valerio, V. (ed.) (2007). *Cartografi veneti: mappe, uomini e istituzioni per l'immagine e il governo del territorio.* Editoriale Programma, Padua.

Winichakul, T. (1994). *Siam Mapped: A History of the Geo-body of a Nation.* University of Hawaii Press, Honolulu.

Wood, D. (1992). *The Power of Maps.* Guilford Press, New York.

Wood, D., Fels, J., Krygier, J. (2010). *Rethinking the Power of Maps.* Guilford Press, New York.

Woodward, D. and Lewis, G.M. (1998). Introduction. In *The History of Cartography*, Woodward, D. and Lewis, G.M. (eds). The University of Chicago Press, Chicago, London.

1

The Map as a Legitimate Space: Cartography as a Language, a Stage and an Issue

Jacques LÉVY
Director, Spatial Intelligence Chair,
Université Polytechnique Hauts-de-France, France

1.1. Introduction

It has been found (Monmonier 2019 [1991]) that the map can be a site of lies. This is often the case in geopolitical propaganda, where states compete with untruths to transform what will be an aggression or a conquest into a defensive posture or a simple recovery of a lost territory. The Nazi map shown in Figure 1.1 is a good example. It was served by the semiological qualities of German cartography at the time which, under the impulse of the Hitlerian cartographer-general Karl Haushofer (1869–1946) and the *Zeitschrift für Geopolitik* journal he founded in 1924, found new balances between data accuracy and communication simplicity.

The reader is easily led to believe that, by their very position, Czechoslovakia and Poland represented serious threats to the Reich. The point can be generalized by saying that, when community visions clash, with allegiance to the nation being one of the most productive in geographical statements, the clash of myths hinders any cognitive approach. Another myth, that of a truth intrinsic to the very idea of the map, collapses: as a language, no more and no less than other languages, the map

For a color version of all the figures in this chapter, see www.iste.co.uk/debarbieux/mapping.zip.

The Politics of Mapping,
coordinated by Bernard DEBARBIEUX and Irène HIRT.

can belong to different discursive registers, that of ideology as much as that of science.

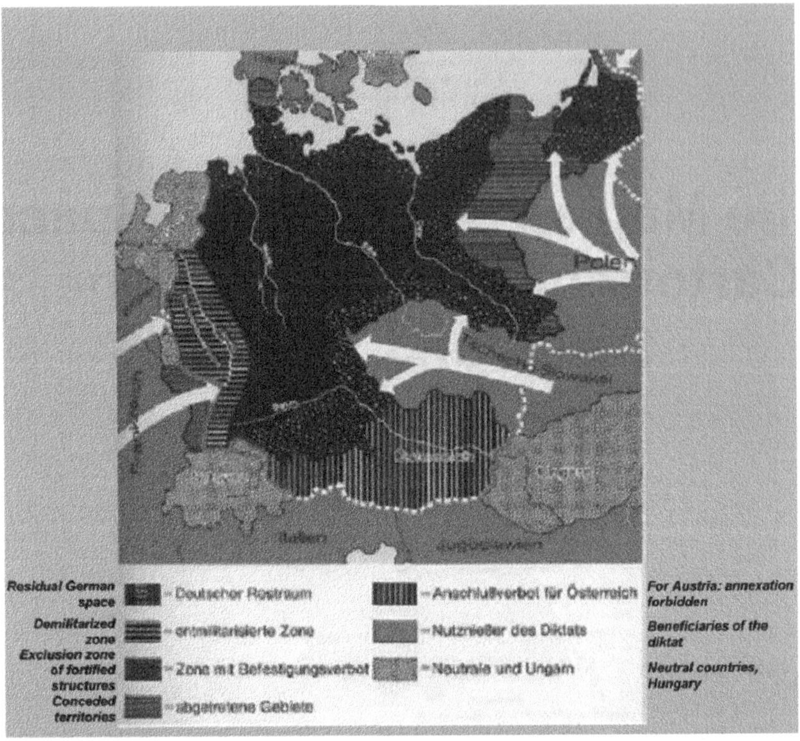

Figure 1.1. *Lying with the map (source: Der Reichsführer SS/SS-Hauptamt, "Der große Betrug von Versailles" in Der Reichsführer-SS, Der Weg zum Reich, Berlin: SS-Hauptamt, 1942, p. 105)*

The use of maps for the purpose of deception does exist and is present in political fights. The question of the regimes of truth that can be identified in cartographic communication deserves to be explored in greater depth, and this will be the subject of the first part of this chapter. However, the relationship between cartography and politics cannot be limited to truth and lies. From this perspective, four other areas of interaction between maps and politics will be addressed: the enrichment of electoral cartography and the uses made of it in public debate; the dynamics of political cartography in relation to recent developments in the social sciences of space ("geographic turn", "spatial turn"); the emergence of an ethical component in the work of the cartographer; and finally, the new ways of representing the relationship between science and citizenship.

As we shall see, these different questions come together in the recognition of the strong presence of the political issues of cartography and, consequently, in the attention paid to preserving and making the cognitive autonomy of cartographic work prosper, which is more essential than ever.

1.2. Map truth regimes

The two maps, as shown in Figures 1.2 and 1.3, of Jewish history have been constructed in an attempt to understand the gap between the dominant view within Judaism and, to a large extent, outside of it, and that of scholars.

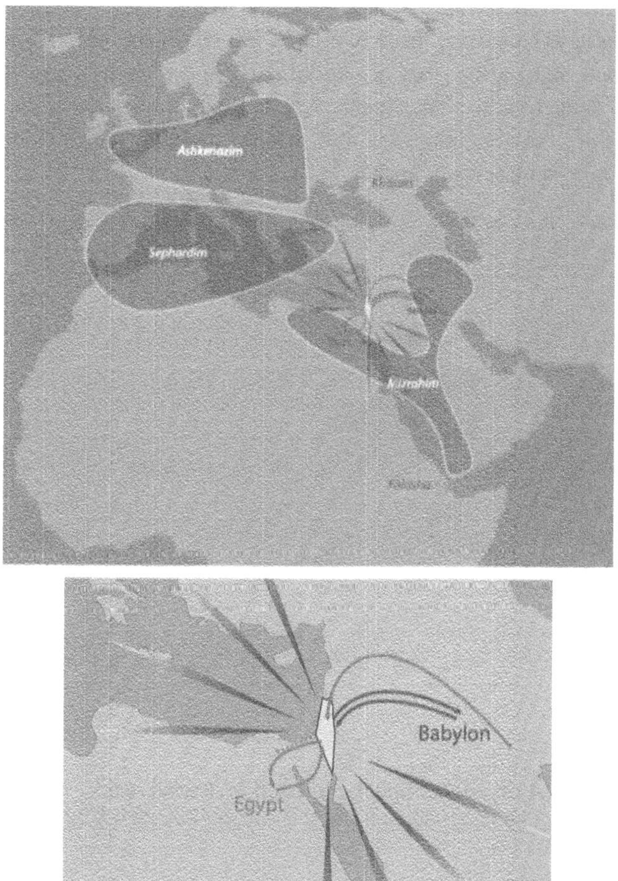

Figure 1.2. *The mythical history of the Jews according to the dominant tradition (source: maps by E. Chavinier and J. Lévy in Lévy 2015)*

In the first map (see Figure 1.2), we are talking about a population located in Palestine, which has been dispersed into many parts of Europe, North Africa and the Middle East, which makes the proposal that this population could rebuild a state in its place of origin through the creation of Israel coherent. In the second map (see Figure 1.3), constructed on the basis of the work by the historian Sand (2008), we discover that the Jews were much more numerous and, at many historical periods, more powerful, outside of Palestine – notably in the Roman Empire, in the Maghreb, in Yemen, in Ethiopia, on the shores of the Black Sea and then in northeastern Europe – than inside; the "Diaspora" never happened and most of the Jews before the Nazi extermination were much more likely to be natives than refugees, and the Palestinians of today are the main descendants of the Hebrews of Ancient Rome. The aim of the project was to make these different visions comparable through graphic semiology. However, these two maps are so divergent in all phases of their conception and realization that it would have been technically acrobatic and intellectually dangerous to make them visually comparable. Here, the map refrains from creating a common language, but instead highlights the incommunicability between two representations of space.

Figure 1.3. *A recent historical approach to Jewish history (-300–1600) (source: map by E. Chavinier and J. Lévy in Lévy (2015, pp. 178–179))*

The notion of a regime of truth, developed by Foucault (1977), can be useful in getting away from simplistic dichotomies. The hypothesis that truth is inscribed in a series of regimes does not aim, against a common belief, for a relativization of truth as such. On the contrary, it helps to show the complexity of the pathways to truth. And truth is ultimately found as the construction of a statement that maintains a clear homology, according to principles of evaluation fixed in advance, with a reality of reference external to this statement. Scientific research imposes a regime of truth on itself, based on coherence (no internal contradiction in the statement), relevance (a search for indicators ensuring an unequivocal link between the theory and the empirical), refutability and communication with all pre-existing or alternative propositions. Esthetics offer other regimes in which the discourse/actual homology is more vague, but not non-existent, and where refutability is replaced by fuzzier modes of validation, leaving room for the subjectivity of the receiver. Art and science can be considered as two cognitive modalities producing truth by being situated on a rational or on an esthetic side (Lévy and Sartoretti 2018). The technical truth regimes that can be summarized by the idea of plan, whether organizational, collective or personal, find their specificity in the fact that the judgment on the correspondence between referent and referred is the achievement of the plan itself. In the affective world, the circulation between the layers of the self (intimacy) produces situations in which the world of sensations and feelings, however autonomous it may be, interacts with the individual's cognitive practices: self-awareness, self-lucidity or hybrids between projects and affective life are common expressions of these interactions. Finally, although the ethical, political and legal worlds do not fundamentally belong to the cognitive sphere and are not intended to produce truth, they are not totally alien to it, because of the strongly argumentative nature of public debates. Thus, a citizen cannot convince another citizen by simply asserting his/her support for one value or public policy over another. They must work to demonstrate that, if they are followed, positive effects will be felt by society as a whole. We can therefore consider that there are specific regimes of truth in the world of affects, interpersonal or ethico-political.

The crude dichotomy between true and false maps is of course still strong when it comes to factual errors or deliberate manipulation of data. For the rest, it appears to be the heir of an era when cartography belonged to the imagination of a weakly reflexive positivist engineering: it was believed that there was one "true map", satisfying the mathematical rules of projection and geolocalization, and all of the others were false. If, on the other hand, we see all maps as true, each in their own way, as can be said, for example, of the discourses of people answering semi-structured interviews as part of scientific surveys, we have a large corpus on which we can propose statements concerning the regime of truth (intentional lying then being, unquestionably, a specific case), but also other aspects, which can unite or separate these different images according to various criteria.

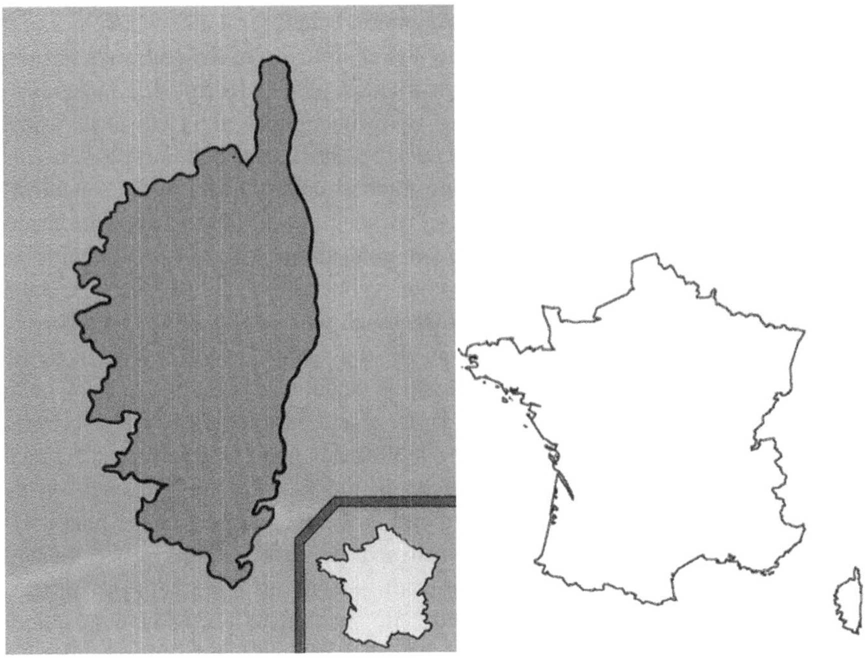

Figure 1.4. *Truths of scale between Corsica and mainland France (source: postcard, Corsica, 1990s; map outline: https://www.rendlemanhome.com/carte-de-la-france-muette.html)*

When we observe a map like this one (see Figure 1.4), it is not so easy to say if it is true or false at first, as the humor which is at the base of its conception plays an essential role and blurs the boundaries of simple classifications. Moreover, this map does not only belong to the universe of technical or scientific truth: it also has ethico-political dimensions (affirmation of the Corsican identity, influence of Corsica in French political life) and esthetic dimensions (effective simplicity of the information economy, effect of the scale inversion). The result is an awareness that, however elementary its design, a map conceals a wealth of meanings that researchers can work with to transform its complexity.

1.3. The new spaces of the electoral map

The above remarks may help in addressing political maps with the required openness. Geopolitics often pushes the notion of truth to its limits (Raffestin *et al.* 1995), and the mapping of inter-state rivalries consists more of a clash of lies, especially concerning the "historical rights" of a particular state over a particular territory, with political maps featuring the competition of truths.

The case of the mutation of electoral spaces in the West allows us to understand this difference. This new configuration, which is omnipresent and massively recurrent, is so clear that it can be formalized simply by means of two equations[1] (Lévy 2020). Its legibility is improved by the use of cartograms. While a Euclidean base map relies on the geometric homology between angles, lengths or surfaces on the space to be represented and on the map, the cartogram opens up the range of possibilities by making the map surfaces correspond to any series of geolocalized variables of the reference reality – population, production, etc. – while respecting a few topological rules (each spatial unit remains well surrounded by its neighbors), which makes it possible to maintain a cartographic relationship between the image and the reality it represents.

Almost everywhere, and with very few exceptions, votes in favor of various types of openness to otherness are clearly more numerous in highly urban areas, those with a powerful combination of density and diversity, such as the centers of large cities, and become weaker as the level of urbanity decreases. This new geographical situation reverses long-standing electoral traditions that were organized more on a regional scale. The shift became visible in the early 1990s, in specific issues, such as the referendums on European construction or the questions of immigration, sexual orientation or religion. It then deepened when the tribune parties that fed on these issues gained influence. Finally, from the early 2000s in the United States, and in the following decade in Europe, it became a generalist marker: the major elections that decide on the major political choices of societies fall within this geographical model. Here, some examples are discussed (see Figures 1.5–1.9).

The first pair (see Figures 1.5 and 1.6), which concerns France, shows that the configuration of the 2017 presidential election was already present, almost identically, in that of the 1992 referendum on the Maastricht Treaty. In both cases,

1 In many recent elections in the West, central areas of large cities have voted in a similar fashion, and so, have peri-urban (or in the case of the United States, suburban and ex-urban) areas. Gradients of urbanity defined by relative position within or outside urban areas (central, suburban/inner city, suburbia/peri-urban, hypo-urban/exurban, suburban) are strongly predictive of these votes. This is not an opposition between the metropolises and the rest of the territory, but a differentiation whose most striking features are the contrasts *within* the large urban areas. This regularity can be expressed in a formalized way through simple equations (see Lévy 2020).

three different scales appear to be relevant: large regions; urban areas, with a strong opposition between cities and the suburban countryside; and, in a more complex way, city centers and part of the suburbs.

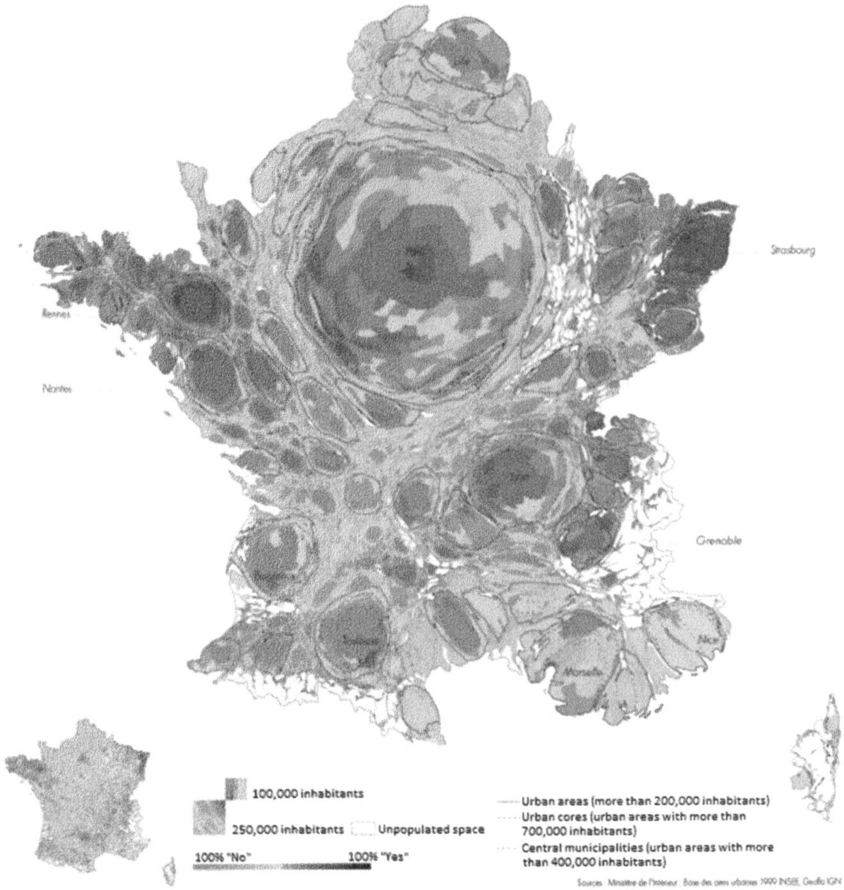

Figure 1.5. *The French referendum of September 20, 1992 on the Maastricht Treaty (source: Lévy* et al. *2017)*

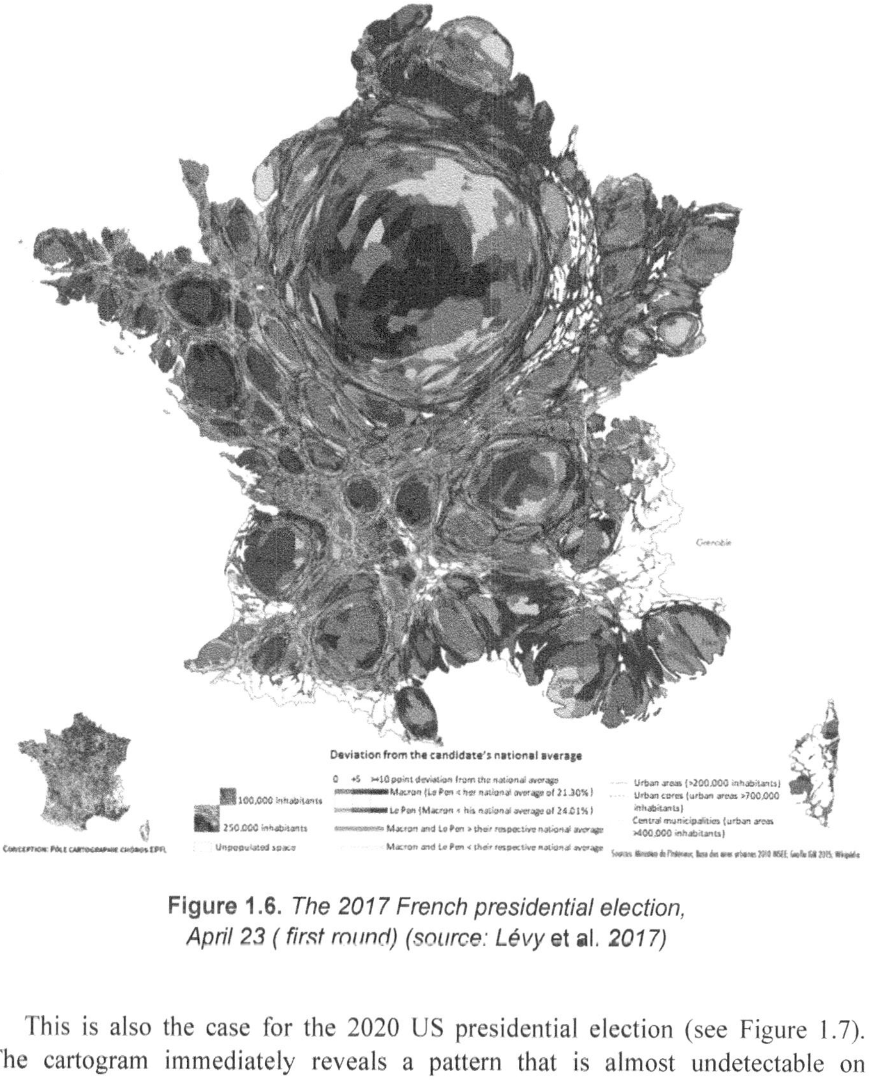

Figure 1.6. *The 2017 French presidential election, April 23 (first round) (source: Lévy* et al. *2017)*

This is also the case for the 2020 US presidential election (see Figure 1.7). The cartogram immediately reveals a pattern that is almost undetectable on Euclidean maps using the same data grid. It is essentially the network of large cities, in whichever state they are located, that voted predominantly Democratic, while a vast continuous swath of large suburbs and countryside opted for the Republicans.

In their construction, these maps seem banal and are comparable to any other electoral map of any place and time. However, it should be noted that the use of the cartogram as a background map, with the population as the reference data, constitutes a semiological extension of the fundamental principle of democracy: "one man, one vote, one pixel", it might be said. More profoundly, these maps express a major change in electoral choices that reflects a bifurcation in the practices of living in the West.

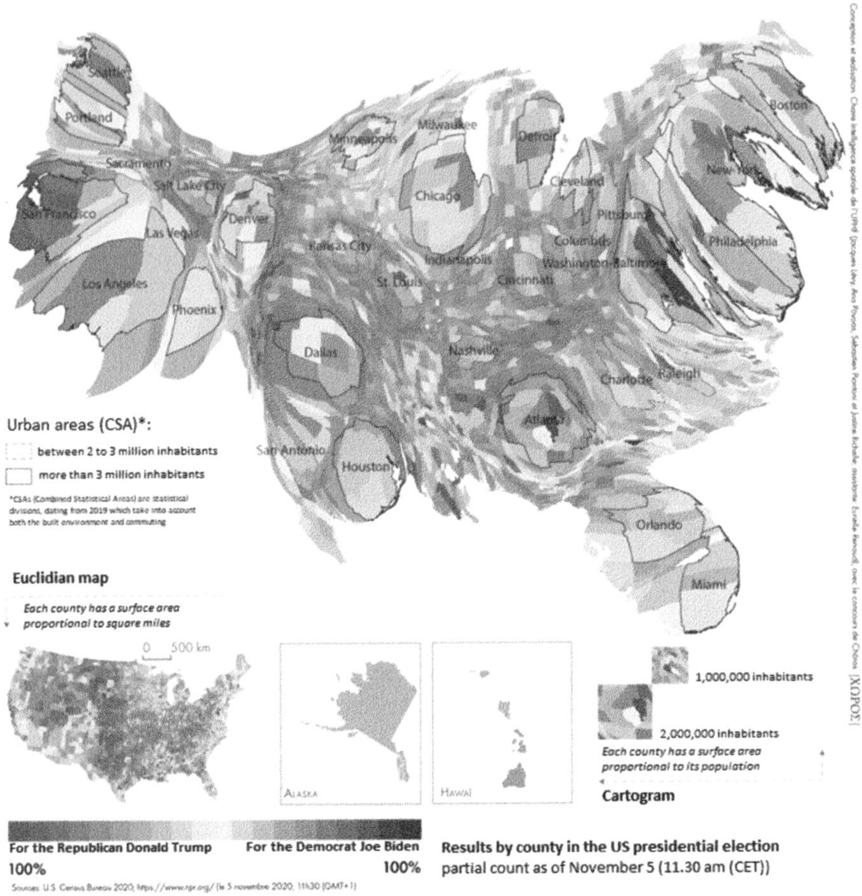

Figure 1.7. *US Presidential Election, November 3, 2020 (source: Lévy* et al. *2020)*

The revolution in mobility allows the majority of the population to make economic-geographical trade-offs involving their place and type of residence, place of work, educational, commercial and leisure practices. The result is that the "residence assignment" – almost a *house arrest* – that heavily marked European rural societies has been greatly reduced and that the place of residence is becoming an indicator of individuals' biographical strategy, without forgetting that the capacity, not only financial but also cultural, to define and carry out such a strategy is not equally distributed in society. This reality, which is both novel and complex, can be usefully compared with another strategic indicator, namely voting in favor of a particular political orientation. It can therefore be said (Lévy *et al.* 2018) that a contemporary electoral map brings two conceptions that can refer to the *right*, face to face. On the one hand, it is about the geographical distribution of political orientations in developed countries, which, almost always argue and decide on issues of justice, through the positions taken by voters on the balance sheets and programs of candidates. On the other hand, the residential choices that determine voter location enable individual actors to apply to themselves, as far as their capacity for action allows, the idea of justice that they deserve to benefit from. These two dimensions come together and interact because the project that each actor implements through their spatiality uses the resources (or, on the contrary, suffers from the absence of these resources) incorporated in an environment that is at least partly the result of public policies.

Finally, what was observed was that these maps were not only representations of other spaces, but of the spaces themselves. In Switzerland, multiple electoral spaces correspond to the responses of voters to multiple referendum questions. These questions are often about openness/closure or identity/otherness, and each time they very clearly oppose the centers of the largest cities to the suburban areas. For a time, however, this reality was masked by the fact that, in Euclidean metrics, cities are not very visible. This allowed the public scene to limit its interpretations to the *Röstigraben* theme, that is, to the country's linguistic divisions, which the Swiss constitutional system largely takes into account and which, unlike the gradients of urbanity, does not threaten sociopolitical or institutional balances.

Gradually, however, new cartographic expressions using the cartogram technique or hybrid variants are making it possible to give greater prominence to densely populated areas. Thus, the following two maps (see Figures 1.8 and 1.9) also present two political spaces in conflict. The first is a common conflict plane in Swiss politics and is explicitly taken into account in the life of the federal institutions; it therefore belongs to the "cruising regime" of the Confederation. The second is new and is *knocking on the door* to be recognized as such.

From the moment it circulates, is received, discussed, and used in public debate, the map – this one and all of the others – does not only account for new political spaces. It creates new spaces that are a part of social life by itself. Indeed, as a representational object that links an external reality and the construction of a discourse on the said reality, the map has a power comparable, in some respects, to the spaces it represents. Thus, when, at the time of major summer tourist mobilities, the public authorities publish a map of expected traffic jams, this is information that is of the same type as that contained in the objective road offer for the territory concerned and which is added to it. The reflexive cognition of travelers will process a first piece of information in the same way: "A highway connects points A and B" – which characterizes the geography of road infrastructures and a second: "The map forecasts that this highway will be saturated and therefore unusable" – which can only be, by definition, discursive. The communicative object thus acquires an active force that puts it on the same level as the pre-existing referent. It is as if we enriched the mapped space with a new geographical layer capable of interacting with the other layers.

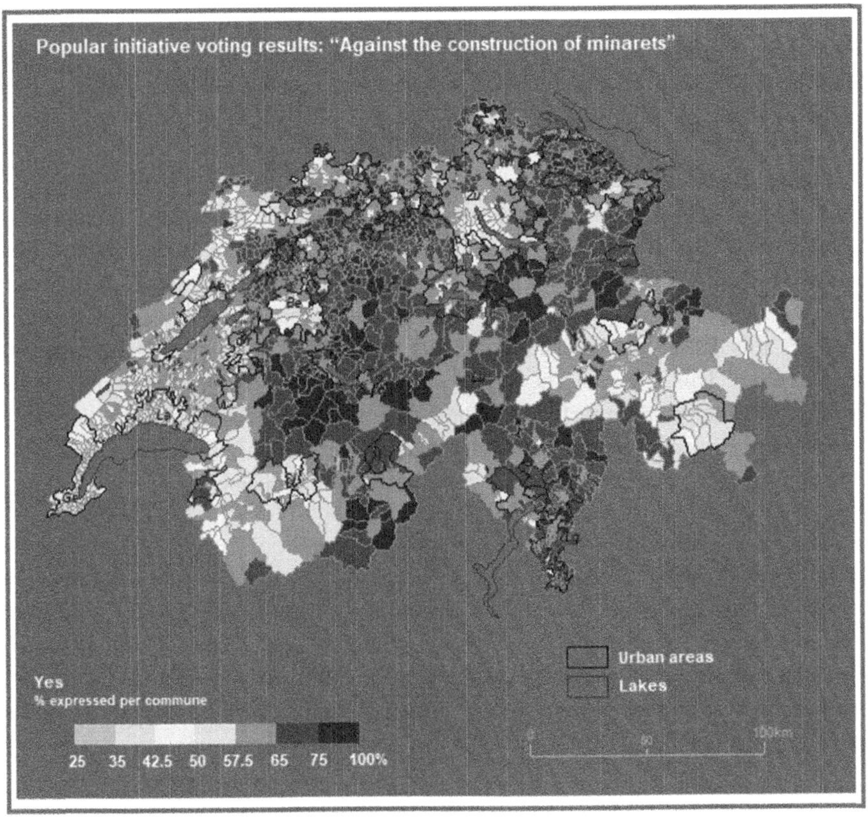

Figure 1.8. *Vote of November 29, 2009 on the ban on the construction of minarets in Switzerland: Euclidean map (source: Chavinier and Lévy 2009)*

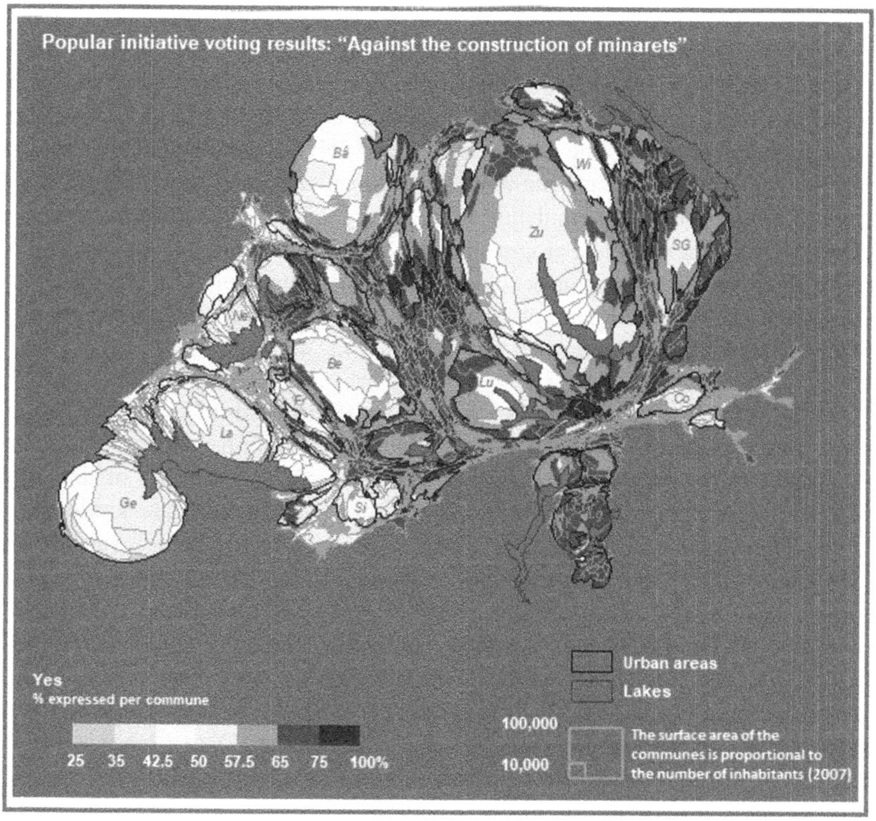

Figure 1.9. *The vote of November 29, 2009 on the ban on the construction of minarets in Switzerland: cartogram (population) (source: Chavinier and Lévy 2009)*

The vote was based on a popular initiative to ban the construction of minarets and was seen by its opponents as a sign of religious intolerance with racist and xenophobic overtones. The vote gave victory to the proposed ban by 57.5% with a turnout of 53.8%.

1.4. Cartographic turn, geographic turn and political space

This awareness of the active force of the "map" object is reinforced and transformed by changes in the context of cartographic production.

Cartography has indeed been turned upside down by the advent of digital technology. In a sense, everything has changed: authors, techniques, audiences and uses. One of the paradoxical effects of these changes is the strengthening of the map as a specific language distinct from all others. Non-geographic mapping (the use of spatial languages to represent something other than space) has been and continues to be influenced by cartography, and has expanded considerably, in turn influencing cartographic semiology. The same can be said for the non-mapping features of satellite geolocation. At the same time, the production of maps is facilitated by the mass of available data, the emulation of the cartographers' guild through the arrival of new players, the cognitive progress resulting from the more explicit consideration of the achievements of graphic semiology and visual communication and the rise of the general public's culture in this field. In the end, the map emerges rather stronger in this rapidly changing landscape.

In this context, the connections of the map with other fields of knowledge become even more significant. Since the map is, from start to finish, a spatial object, the *geographic turn* in the social sciences (taking the spatial dimension of the social seriously) and the *spatial turn* in geography (new theoretical developments in thinking about this dimension), which have been taking place over the last half-century, offer new resources and requirements for the cartographer's work. Among the many innovations that have resulted, we can note the success, less and less contested, of the Leibnizian conception of space (Harvey 1969; Werlen 1992; Lévy 1994): the Newtonian model of the prior empty frame that the Newtonian approach carried has receded. On the contrary, the idea tends to impose that there is no space in itself but also, as proposed by Leibniz (Robinet 1957), that space is nothing else but the set of distance relations, according to various metrics, between localized realities. These realities are not *in* space, but they *make* space, and this is not without consequence for the notion of the background map, which ceases to be a tool of naturalization (in both senses: dehistoricization and naturalism), and becomes a sort of summary of all of the pre-existing maps that the one we are drawing deserves to be compared to. This cartographic counterpart of geographical *extent* makes each map an open dialogue between a background and a theme. From this perspective, *spatialities* – the actions of humans in the geographical environments that are *spaces* – can take their full place, and will no longer be seen as second or secondary. Thus, the definition of urban areas includes the mobility of inhabitants as a primary datum, which means that they contribute through their practices to shaping the environment that nevertheless encompasses them. The strength of spatialities as a political issue is reflected in a multiplication of scales, from local to global, and of metrics, with a pre-eminence of networks and their multiple interactions.

A map can then be used as a tool to understand the relationships between the *prompts* issued by the environments and the *affordances* expected by the actors. The notion of affordance was proposed by Gibson (1977, pp. 67–82) to account for the resources that the user of a space finds and uses to enable them to carry out a project: this can be an inclined plane for a suitcase or a pushchair, an information board, a bus stop, people who can be questioned, or the urban legibility ensured by a simple map or recognizable monuments, in short, any element that makes it easier to use a place. A *prompt* is an incitement to action proposed by spatial environments (Rudler 2018). It is close to behavior settings, a notion that Barker (1968) uses to describe devices that, by their configuration, send messages, more or less imperious, to those who approach them that indicate that a specific behavior is expected of them. More generally, the prompt applies to any situation in which it is the environment, arranged or not, that proposes an action to the actor, and this at all scales: it can be the layout of a station platform or a photo of a beach bordered by a turquoise lagoon. In the encounter between prompts and affordances (Koseki *et al.* 2020), the construction of a spatial capital is at stake, a combination of experiences and skills that gives an actor or a place the capacity to act on geographical environments (Lévy 1994).

From this perspective, inhabitation can be seen as the spatial component of *resonance*, if we wish to use Rosa's (2019) expression. If we define inhabitation as a dialogical interaction between actors and environments (Lazzarotti 2006; Frelat-Kahn and Lazzarotti 2012) or, more precisely, between spatialities and spaces, this creates a tension due to, in Rosa's words, the relative "unavailability" of the external world to actors' projects.

From a cartographic point of view, this relationship between actors and environments has long been tilted in favor of spaces. It was up to the user to cope with maps that provided reliable information, but were not designed to facilitate their action. In recent years, however, there have been innovative alternatives in this area. Transport for London has placed maps of the area at the exit of underground stations and these maps are "anthropocentric", that is to say, their orientation does not place north at the top (as is the case with "geocentric" maps), but instead what is in front of the pedestrian, which allows the latter to use this affordance to find their way more easily in an unknown space. In retrospect, there is a realization that before these maps were introduced, the passenger exiting the underground was expected to find their way on their own, until they found a map, sometimes very distant and geocentric.

The relationship is thus rebalanced by the fact that, on the side of spatialities, putting actors on the map is now easier due to digital technology (Calbérac *et al.* 2019). What was once confined to the statistical imagination and its relentless pursuit of the average man is greatly enriched by the appearance of individualities that are no longer sums of factors or propensities, but complex autonomous systems. We tend to have as much data, mostly localized, for an individual as we used to have for a whole society. We must learn to work together with mass data and individuals, and tame these powerful singular spaces that maps can also become. The risk that they will be used, like other individual databases, for purposes of control or manipulation is not negligible. However, this new landscape also undoubtedly gives new margins of freedom to the citizen-inhabitants, which means new responsibilities for them at the same time. On these innumerable maps populated with *I*s, how do *we* construct the politics? Undoubtedly not by the magic of cartographic language alone, but rather by the processes that assemble citizens around their map(s).

1.5. Ethical turn and mapping

Thus, it is not only as inhabitants that people "get on the map", but also as citizens. For a long time, cartographic controversies superbly ignored them. The truth/lie pairing applied to maps was situated within a framework of enunciation that made the cartographer the holder of an intrinsic truth linked to their technical command, making their relationship with citizens a master/student one at best.

On the contrary, it may be thought that if researchers and cartographic engineers have indeed signed a contract with society consisting of concern for the truth of the maps they draw, this does not *ipso facto* guarantee that they are immune against any slippage in terms of the truth regime. They are specific actors, but actors among others, who do not, for example, have a monopoly on argumentative competence. Arguing is what citizens do in public debate, while the world of research must accept the tension that fundamentally characterizes it, between an abundant and difficult-to-decipher empirical material without prior reference points and a theory that is always threatened by dogmatism or immobilism. Where social scientists can make a useful difference is in their skills of "reconstruction" in the sense of Ferry (1991): *thinking together,* the plurality of practices and registers of discourse – narrative, interpretive and argumentative – produced by the various actors present in society. Thinking together does not mean reducing divergences or denying conflicts. Rather, it means seeking translations that allow each person to perceive, identify and discuss the other's point of view. This is not self-evident and it is not by constructing a list, a sum or a point of cancellation of forces that the solution can be found, because what one risks obtaining then is a neutralization, a depoliticization of the debate. In contrast to the illusion, both literal and figurative, of the existence of a hidden barycenter (a midpoint where the gaps between the stakeholders would become as small as possible), the project of mapping the public debate consists first of describing these gaps and then seeking another spatiality (other scales, other metrics) that would make it possible to either overcome or settle the conflicts. In other words, we must not hope to find a resolution to conflicts either in a prior consensus that would have been hidden by the confrontation (this is imaginable but unlikely), or in a greater common denominator that would force everyone to come down (this would be a "bottom-up" solution, which is certainly sometimes necessary, but only as a last resort) – all approaches that result in poor cartography, consisting of adjusting locations on an unchanged background map. The challenge of citizen cartography is quite different: to co-design a map which, in terms of background and theme, does not yet exist. It is through this staging, whose plot and epilogue ultimately belong to the citizens, that the cartographer can contribute by spatializing the actual dissensions and the conceivable consensuses.

To move in this direction, it is easy to understand that it is not by virtue of a moral leap of faith (Kant 1798, pp. 203–221) that we can move forward. It is rather a question of ethics (Spinoza 1954 [1677]; Kant 1785; Ricoeur 1990; Lévy 2021), as a compatibility of the logics of society and those of its components. It is a historical construction of values in motion by actors united by a continuous dialogue (Lévy *et al.* 2018). We can call the tilting point – which is in fact a slow process – an *ethical turn*, from which the world of transcendent injunctions, antinomies and exceptions that constitutes the moral universe gives way to the societal co-construction of values proposed and discussed by all actors.

Table 1.1 shows the differences between a moralistic approach and an ethical approach. The moralist approach is both overbearing and revealing, and consists of denouncing, through technical mastery and good intentions, the misuse of maps. Conversely, the ethical approach is much more demanding because it no longer places the cartographer in a dominant position, but as one of the players in a world that is moving in many ways. On the supply side, the possible themes and the way maps are made are changing. On the demand side, society's expectations, the skills and practices of map consumers, and, for all of them, society's ethical values, which have probably changed more in a century than in the whole of human history, are changing. Thus, for all actors and precisely because they are actors and not only agents, the association between freedom and responsibility becomes cardinal. Similarly, equality only makes sense if it translates into access to effective capabilities for all (Sen 2010), which means that all education must be based on an obligation to achieve results and not only on means. These elements also concern the relationship with the map, and it is understandable why cartographers cannot "give lessons" to other stakeholders in the sector.

In this framework, the properly political dimension cannot be reduced to its institutional, sociological or psychological components. Justice, which is the permanent central issue of political deliberation, is neither care, nor simple reciprocity in relations between two people, nor even negotiation between groups that do not share common values (Lévy *et al.* 2018). Justice is the contradictory synthetic summary of what citizens say about what is just. And if we want, moreover, to map the idea of justice, we have to go looking for what is really at stake: the equitable public redistribution of private goods, but increasingly, in contemporary societies, the co-production of public goods, such as, on the one hand, education, health, culture, nature, which have a powerful spatial component, and, on the other hand, urbanity and mobility, which are spatial realities throughout.

	Moral	Ethical
Sociality/society relationship	Antinomy	Compatibility, convergence
Principle	Injunction	Value
Status of statements	Transcendent	Self-organized
Creation of new statements	Dogmatic (revealed substance)	Pragmatic (substantial constructed + procedural approach)
Epistemology	General/personal	Singular/universal
Relationship with the truth	Concealment/revealment	Deconstruction/reconstruction
Public relations	Cognitive technocracy "I know, you don't."	Reflective explanation "I know, you know, we think."
Social values	Domination, egalitarianism, charity. Freedom/equality antinomy	Fairness, solidarity, responsibility. Compatibility of freedom and equality
Relationship with universality	Proclaimed but ineffective universality	Historically constructed universality

Table 1.1. *The ethical turn (source: adapted from Lévy 2015)*

1.6. Mapping, citizen science

All of these elements result in much richer interactions between cartographers and the rest of society. Through its formalism, the map provides a specific perspective that influences the public debate.

For example, anyone can consult the maps of the Census Bureau, a federal government agency, which show the slow crumbling of New York ghettos, including those in Harlem, the East Village and the Lower East Side (Lévy 2014). Yet, an academic-political current had defined any change in the social composition of these neighborhoods as "gentrification" and a revenge of big capital. Today, we can see that it is the mix that has progressed, with the combination of socioethnic and socioeconomic groups approaching (but not yet reaching) the New York metropolitan area. This clarifies the debate: those who denounce *gentrification* in the wake of a school of thought ranging from Smith (Smith and Williams 1986) to Soja (2010), actually fear mixing because they believe that the identity of the poor must be preserved at all costs due to homogeneous inhabiting arrangements. This is a defensible political choice, but one that was obscured by the claim that increased mixing was simply impossible (Harvey 1973). Here, the map helps to make crucial experiments that can encourage ordinary citizens to enter the political arena possible and, according to the values, judgments and options they defend, propose commensurations between emergencies that make it possible to construct coherent public policies. In US cities, the debate between traditional policies that lead to the separation of social groups and attempts to promote a more socially mixed society is nourished by multiple maps, as was the case in the conflicts over public transport in Los Angeles and San Francisco (Bénit-Gbaffou 2005; Bénit-Gbaffou *et al.* 2007; Soja 2010). The cartographer cannot claim to short-circuit this process by providing knowledge that would render the debate useless, but, on the contrary, can contribute to the freedom of deliberation by proposing an explanation that makes it possible to keep misunderstandings and false leads at bay. By being present on the map and talking to each other through the map, ordinary citizens are therefore the most essential emerging actors.

"Participatory mapping" has yet to be invented as a component of interactive democracy, which complements rather than challenges representative democracy. Where representative democracy is still in its infancy, the project of replacing it by direct interaction with inhabitants alone is hardly tenable. Up until now, those who design, draw, or interpret maps have sometimes affirmed their willingness to promote the *empowerment* of inhabitants, but on the condition that they endorse their own ideas, which are sometimes marked by the project of preserving the "authenticity" of "cultures" or the integrity of "civilizations" in the face of powers that do not care about them. Why not? – if that is what the people concerned are asking for, and if they are also entitled to distance themselves from the founding myths of their own society and refuse allegiance to community powers. In other words, on all of these subjects, it is clearly not up to the cartographer to decide for the inhabitants what is good for them, but rather to respond, as honestly as possible, to their demands or expectations, which are, more often than not, marked by the dissensus inherent in any political life. The territorialization into "countries", by continuous and bounded metrics of "indigenous" (pre-colonial) spaces that were mostly reticular in nature, in Australia, North and South America (Hirt 2012), or Arctic societies (Collignon 2002), participates in the projection of a classical state model (Elden 2013; Debarbieux 2015) onto societies (or the heritage of societies) without a state. Once the necessary (apologies, reparation, recognition, equalization of rights and status) has been done so that the past can "pass by", it is the combination between two modes of inhabitation, the mental space of memory and the space practiced today, the heritage and the plan, that becomes the political issue for today. The map can then constitute a link (Desbiens *et al.* 2017), but this presupposes a posture of the cartographer, less carried by militant certainties and more attentive to the complexity of situations.

What do citizens expect from cartographic *experts*? What we are increasingly seeing is that citizens are asking for training that empowers them to play a role in urban planning or development issues, and, more generally, spatial identity and development. They do not want to receive a lesson or an ideological message, but want to be supported in a process of expanding their capabilities (Sen 2010; Lévy *et al.* 2018).

In Medellín (see Figure 1.10), members of a neighborhood committee are asking academics to teach them how to produce and read maps, so that they can discuss the issues as equals with the local government, urbanists and other actors in urban society. To be fully effective, *mapmakers* must unapologetically abandon their engineering role and construct another, which consists not only of making themselves available, but also in adjusting their technicality (and especially their language) to the specificities of the issues and political scenes. The pluralism intrinsic to all political life also requires the plurality of cartographic projects and not just the variation of locations on a standardized base map. The commensuration of all of these citizen cartographies is not the point of departure, but the point of arrival, to be invented, of the public debate.

Figure 1.10. *Citizen Mapping in Medellín, 2016 (source: original photograph by Jacques Lévy)*

Like all political mediators, cartographers must accept the fact that they are actors in a world of actors. They are only actors among others, many others make maps and corporatist defenses are no longer appropriate. Maps circulate in a horizontal universe where no one can afford to address others as if they are in a dominant position.

In the old system, as it existed in the West in the 19th and 20th centuries, the pairing of cartographers with those of political or economic power dominated the scene. Innovations were emerging, such as the gradual emergence of a "thematic cartography", which took the map out of the "headquarters", but two significant limitations constrained the creativity of the field. First, a positivist epistemological paradigm gave the illusion that drawing a map was only about reproducing an already existing reality, and not about choosing a point of view and defining a method. This led to an impoverishment of languages and a belief that the Earth and the World only needed Euclidean geometry to be represented. Second, the concentration of cartographers in "countries", those territories that are continuous inwardly and bounded outwardly, which constituted the privileged horizon of their "clients", delayed the realization that networks constitute both the basic fabric of spaces and the fundamental spring of spatialities. This willful blindness has even diverted, until recently, the curiosity of specialists from the interest they should have shown in the remarkable achievements of Arab-Persian cartography from a thousand years ago. The impressive history of cartography launched by Harley and Woodward (1987–2015) has fortunately helped to show the extraordinary richness of the world's cartographic culture. It was necessary to wait for the instructions for use, a geography inserted in a renewed science of the social capable of linking spatial productions, of which maps are obviously a part, and increasingly massively thanks to digital technology, to other human productions.

Ordinary citizens have been invited to the map table. Like everyone else, they need to acquire a minimum amount of technical skills to have access to the language of maps, but their arrival cannot be said to have "lowered the level" of communication in this field: quite the opposite, in fact. They also bring a breath of fresh air because they come with questions and are, moreover, quite capable of arguing if answers are proposed to them.

For all of these reasons, cartographic engineering is changing in nature and status. As can be observed in another field of spatial "technology", urban action, cartographers and urban planners are no longer "experts" with all of the answers, but rather mediators with specific cognitive skills between inhabitants, habitability, for which politicians, including citizens, are responsible, and inhabitation, which society implements.

The relations between maps and politics are multiple. In this chapter, the themes of geopolitical propaganda, spatial ideologies, electoral geography, urban planning and citizen democracy show the multiplicity of the subject. In all cases, they raise the question of the specific role of the cartographer, whether they are a technician or a researcher. They must define their own place as a producer of knowledge without taking the place of citizens. This distinction in turn opens up a debate common to all on the ethics of knowledge and communication. The cartographer is not only someone through whom information passes, they are also the one who produces and disseminates it, which places on their shoulders, a responsibility commensurate with the questions and answers of which the act of drawing a map is an integral part. The cartographer's contribution may sometimes seem modest, but it should never be underestimated because it is a part of the long chain of production, interpretation and action, of which the map is an increasingly present link.

While reinforcing the specificity and richness of its language, cartography has become established in the political scene and now plays a major role in the emergence and dynamics of its issues. A new, more open and more demanding age is therefore upon us.

1.7. References

Barker, R.G. (1968). *Ecological Psychology: Concepts and Methods for Studying the Environment of Human Behavior*. Stanford University Press, Stanford.

Bénit-Gbaffou, C. (2005). Politique des transports collectifs et démocratie locale à Los Angeles : entre participation et fragmentation. *Flux*, 2, 6–22.

Bénit-Gbaffou, C., Fol, S., Pflieger, G. (2007). Le front anti-métro en Californie. Controverses autour des politiques de transport en commun. *L'Espace geographique*, 36(2), 115–130.

Calbérac, Y., Lazzarotti, O., Lévy, J., Lussault, M. (2019). *Carte d'identités. L'espace au singulier*. Hermann, Paris.

Chavinier, E. and Lévy, J. (2009). Minarets : malaise dans l'alteridentité. *EspacesTemps.net* [Online]. Available at: https://www.espacestemps.net/articles/minarets-malaise-dans-alteridentite.

Collignon, B. (2002). Les toponymes inuit, mémoire du territoire. Étude de l'histoire des Inuinnait. *Anthropologie et Sociétés*, 26(2–3), 45–69.

Debarbieux, B. (2015). *L'espace de l'imaginaire. Essais et détours*. Éditions du CNRS, Paris.

Desbiens, C., Rivard, É., Hirt, I. (2017). Nous nous souvenons du territoire : la géographie québécoise face aux mémoires autochtones. *Cahiers de géographie du Québec*, 61(173), 293–313.

Elden, S. (2013). *The Birth of Territory*. University of Chicago Press, Chicago.

Ferry, J. M. (1991). *Les puissances de l'expérience : le sujet et le verbe*. Le Cerf, Paris.

Foucault, M. (1977). Vérité et pouvoir : entretien avec Fontana, M. *L'Arc*, 70, 16–26.

Frelat-Kahn, B. and Lazzarotti, O. (2012). *Habiter : vers un nouveau concept*. Armand Colin, Paris.

Gibson, J.J. (1977). The theory of affordances. In *Perceiving, Acting, and Knowing: Toward an Ecological Psychology*, Shaw, R. and Bransford, J. (eds). Lawrence Erlbaum Associates Inc, Hillsdale.

Harley, B. and Woodward, D. (1987–2015). *The History of Cartography*. Chicago University Press, Chicago.

Harvey, D. (1969). *Explanation in Geography*. Edward Arnold, London.

Harvey, D. (1973). *Social Justice and the City*. Edward Arnold, London.

Hirt, I. (2012). Mapping dreams/dreaming maps: Bridging Indigenous and Western geographical knowledge. *Cartographica*, 47(2), 105–120.

Kant, E. (1785). *Fondements de la métaphysique des mœurs* [Online]. Available at: http://classiques.uqac.ca/classiques/kant_emmanuel/fondements_meta_moeurs/fondements.html.

Kant, E. (1798). Le conflit des facultés. In *Opuscules sur l'histoire*. Garnier-Flammarion, Paris.

Koseki, S.A., Lévy, J., Sartoretti, I. (2020). De l'espace pour l'école de la réussite. *Zenodo* [Online]. Available at: http://doi.org/10.5281/zenodo.3773955.

Lazzarotti, O. (2006). *Habiter, la condition géographique*. Belin, Paris.

Lévy, J. (1994). *L'espace légitime*. Presses de Sciences Po, Paris.

Lévy, J. (2014). Science + space + society: Urbanity and the risk of methodological communalism in social sciences of space. *Geographica Helvetica*, 69(2), 99–114.

Lévy, J. (ed.) (2015). *A Cartographic Turn: Mapping and the Spatial Challenge in Social Sciences*. EPFL Press, Lausanne.

Lévy, J. (2020). Ce dont on peut parler, il faut le dire. *EspacesTemps.net* [Online]. Available at: https://www.espacestemps.net/articles/ce-dont-on-peut-parler-il-faut-le-dire.

Lévy, J. and Chavinier, E. (2009). Malaise dans l'alteridentité. *EspacesTemps.net* [Online]. Available at: https://www.espacestemps.net/articles/minarets-malaise-dans-alteridentite/.

Lévy, J. and Sartoretti, I. (2018). Arts et sciences : le temps de l'hybridation. *EspacesTemps.net* [Online]. Available at: https://www.espacestemps.net/articles/arts-sciences-le-temps-de-lhybridation/.

Lévy, J., Poncet, P., Tricoire, E. (2004). *La carte, enjeu contemporain*. La Documentation française, Paris.

Lévy, J., Maitre, O., Póvoas, A., Fauchille, J-N. (2017). *Atlas politique de la France. Les révolutions silencieuses de la société française*. Éditions Autrement, Paris.

Lévy, J., Fauchille, J-N., Póvoas, A. (2018). *Théorie de la justice spatiale. Géographies du juste et de l'injuste*. Odile Jacob, Paris.

Lévy, J., Piantoni, S., Póvoas, A., Renaud, E., Richelle, J., Schmitt, G. (2020). Topologies politiques. Espace politique et dynamique sociale après l'élection présidentielle américaine de 2020. *EspacesTemps.net* [Online]. Available at: https://www.espacestemps.net/articles/topologies-politiques/.

Monmonier, M. (2019). *Comment faire mentir les cartes*. Éditions Autrement, Paris [Original published 1991].

Raffestin, C., Lopreno, D., Pasteur, Y. (1995). *Géopolitique et histoire*. Payot, Lausanne-Paris.

Ricœur, P. (1990). *Soi-même comme un autre*. Le Seuil, Paris.

Robinet, A. (1957). *Correspondance Leibniz-Clarke*. PUF, Paris.

Rosa, H. (2019). *Résonance. Une sociologie de la relation au monde*. La Découverte, Paris.

Rudler, J. (2018). L'urbanisme de l'entre-deux. PhD Thesis, EPFL, Lausanne.

Sand, S. (2008). *Comment le peuple juif fut inventé*. Flammarion, Paris.

Sen, A. (2010). *L'idée de justice*. Flammarion, Paris.

Smith, N. and Williams, P. (1986). *Gentrification of the City*. George Allen and Unwin, London.

Soja, E. (2010). *Seeking Spatial Justice*. University of Minnesota Press, Minneapolis.

Spinoza, B. (1954). *Éthique*. Gallimard, Paris [Original published 1677].

Werlen, B. (1992). *Society, Action, and Space*. Routledge, London.

2

Cartography and Spatial Production of Society

Franco FARINELLI
University of Bologna, Italy

2.1. The map, politics and morality

Behind its sophisticated and scholarly trappings, the bourgeois geography that emerged in the 18th century hid a terrible ruse behind the apparent innocence of its simple forms. The name it was given in the Germanic world – *Reine Geographie*[1] – was in itself a fatal deception, for it served only to conceal its true intention. German historiography, the only subject to have preserved the memory of its birth and institutionalized its name, has retained 1726 as an emblematic date; it is the year of publication, in Helmstedt and in Latin, of the *Commentatio de Vera Geographiae Methodo* by Polycarp Leyser. According to Emil Wisotzki, this work "suddenly made geography something completely different" (Wisotzki 1897, p. 197). This was not because of the originality of its content, but because it transformed what already existed in geography as an aphorism or embryo into a specific critique of the "purely political representation" of the earth's surface. By political representation we mean a representation that, according to Leyser himself, "considers the political character of states, that is, the distribution of lands and localities within the state perimeter" (Leyser 1726, pp. 5–6), according to the political aims (with an emphasis on boundaries and borders: *fines*) of those who hold power. In contrast, the "true geography" advocated by Leyser leaves such concerns to statesmen, and is instead about "all those variations of the ecumene (*orbis habitatus*) imposed and constituted

1 Literally "pure geography" to be understood as "geography in its pure state".

The Politics of Mapping,
coordinated by Bernard DEBARBIEUX and Irène HIRT. © ISTE Ltd 2022.

by Nature itself" (Leyser 1726); basically, it is primarily to do with natural divisions. It therefore takes the name of "natural geography" or "pure geography"; the purity of which lies in the rejection of the principle of the utility of geographical knowledge itself, in other words its submission to any function of a political nature. It was precisely with this last name that the first form of bourgeois geography had its own fate: geography for geography's sake, the claim to neutrality of geographical knowledge, the claim to its right to be useless, in appearance. But in reality it was much more than that.

How is it possible to classify the different parts of the earth's space? How can the inhabited planet be divided? What "order" should be assigned to the description of the different parts of the known world? To what logic should the "geographical system" be subjected, and how should it be composed? Here are the first questions of bourgeois geography, which deliberately places it at odds with the aristocratic geographical knowledge of feudal origin. For Leyser, neither of the two orders of discourse seems satisfactory. The first, the alphabetical order of topographical lexicons and toponymic inventories, "does not suit any true discipline", any "true science". The second, the political order, based "on the origin of populations or on the political borders of states," is "unstable" because wars and peace treaties continually change the state of things. Leyser, therefore, proposes a third type of order; he suggests a different system, based on the "natural position" and "situation" of cities in relation to "natural subdivisions", in relation to "mountains, valleys, springs, rivers, lakes, seas, in a word, to all the various things present on the inhabited surface of the Earth and by Nature as it is worked out and situated" (Leyser 1726, p. 5). And this is because "these signs are the most stable, for in fact it is not very easy to remove mountains or divert the course of rivers or turn the sea into dry land; and even if something like that happened, historians would not pass it over in silence" (Leyser 1726, pp. 5–6).

The immediate response of the state geographers, the *Staatsgeographen*, proves that Leyser's proposal, which was expressly rejected, was a real political critique, not of geography, but *through* geography. It also only had one immediate goal: to try to undermine the hegemony of the cartographic dictatorship over geographical knowledge. Gustav Conrad Hering, the Prussian Imperial War Councillor and author of *Pensées sur l'utilité et la nécessité de la géographie* (translation: Thoughts on the Necessary Utility of Geography), summed up what was at stake in Leyser's proposal: "From time immemorial, political subdivisions and the names (*Benennungen*) derived from them have been the main rule. But for the so-called *geographia naturalis* this is no longer the case" (Hering 1728, pp. 47–48). For Hering, on the contrary, the naming of space must remain "indissolubly linked" to (i.e. dependent on) the political subdivision, and the order of contents in geography books strictly reflects the hierarchy of power: "first the largest portions of (political) space, then the smallest" (Hering 1728, p. 27). The equally famous Eberhard David Hauber (1727) asks: What should we call the natural areas that Leyser claims exist? According to him, "natural geography" is certainly not able to provide the answer, because only the politically delimited space has a name: that of the state and its institutions. Therefore, Hauber concludes, the political partition must remain the foundation of geographical analysis; moreover, one must start with it because "most readers are more disposed towards this partition than towards the others, and it is therefore much more important. Otherwise, what a mess would result from a reversed order!" (Hauber 1727, p. 71).

For state geographers, cartographic representation, in addition to identifying itself with reality in an immediate and absolute manner, decides, precisely by virtue of this identification, what exists and what does not, and therefore what is legitimate to think about. However, once the ontological, politically oriented function of the cartographic image has been admitted, the controversy moves from the explicit to the implicit. And what is not said, precisely because it is the inevitable and obvious consequence of this admission, guides the very intention of Leyser's proposal, its motivation: since naming, like the cartographic production that sanctions its validity, is an act that falls to political power, claiming to name space in a different way in fact denotes a desire to produce maps that are different from those that already exist, that is to say, it is claiming to establish over space itself a different power, a hold of a different nature from the one that exists.

The new geography thus transposes onto the level of the conception of geographical space a claim that is the opposite of the form, expressly and immediately political, by which it is then accounted for; it orients and leads to the concrete control of space even before that of its representation. The "disorder" invoked by Hauber in his final comment must be understood literally. Criticism of the order of geographical discourse is in fact equivalent to criticism of the existing social order, and the old school of state geographers is fully aware of this. Theorizing, as Leyser does, a knowledge that is no longer subject to the point of view of political utility only serves to try to free it from political power in order to subordinate it to another kind of politics, to put it at the service of another form of power. Only in the name of its own imaginary purity can "natural geography" subject the geography of the state to its own judgment. Moreover, this polarization between "natural geography" and state geography reflects another polarization, equally important in the 18th century: that between morality and politics. Reinhardt Koselleck has explained that bourgeois political criticism does not consist of moral judgment as such, but is guided, at least initially, by a concern to distinguish morality and politics. Likewise, it can be said that the dualistic division of geographical knowledge into natural and political domains is in its historicity the premise and consequence of the political critique of geography. Like moral judgment, geographic judgment "becomes political criticism not only insofar as it issues its own harsh verdict on politics, but also conversely, insofar as it is an instance of judgment outside the political sphere" (Koselleck 1979 [1959], p. 194). And both moral judgment and geographical judgment obey the strategy of the *Weltbürgerplan*, the bourgeois plan for world domination which Koselleck says aims to achieve an "indirect control of power" (Koselleck 1979 [1959], pp. 118–119): to detach oneself from the absolute state – a prerequisite for the establishment of a civil society – "first of all in order to escape its influence, but also in order to invest the state in an apparently apolitical way precisely on the basis of this detachment" (Koselleck 1979 [1959], p. 194). What interest us specifically here is that the conflict between defenders of the old order and promoters of the new order is crystallized less on the territory than on its map if, in the end, the whole bourgeois strategy, the whole *Weltbürgerplan*, were cartographic in nature.

2.2. The map, territory and globe

For centuries, there was no difference between a map and a territory; this homology was the very foundation of what we mean by "state" today. Only in the writings of Nicolaus Machiavelli and the writers of the Italian Renaissance did the term begin to take on the meaning of domain, territorial possession or object, although this meaning still coexists with another, older one, which refers to the notions of authority, domination and the political power of the prince (Chabod 1961, pp. 143–174). It was only in the 16th century that the state was transformed from a function to an actual *thing*; this process of objectification is quite similar (including its cartographic determination) to that which took place between the 16th and 18th centuries for the idea of the city: at first denoting a group of men and women, a complex of relations, a style of relationships, but later became the opposite, a set of buildings, of material things, fixed, within reach; the commodification of these elements presupposing their chosification (Farinelli 2003, pp. 136–137; see also Farinelli 2009).

It is this process that Kant defined in the preface to the second edition of the *Critique of Pure Reason* as the principle on which all modern science rests: "reason has insight only into what it itself produces according to its own design (*Entwurf*)" (Kant 1998 [1781], p. 109). Let us recall that *Entwurf* means project, but also, more precisely, a sketch or a drawing, which corresponds to reality only by approximation. And it is, once again, the same process that Carl Schmitt referred to with his use of the expression "*mise en forme*", used to designate the abandonment between the 16th and 18th centuries of the medieval conception of war and the related abandonment of the concept of a "just cause" (Schmitt 2001 [1950]). The process is the same because it arises from a single act, from what Jean Baudrillard, mistakenly attributing it to the beginning of the postmodern era, calls the "precession of the simulacrum" (Baudrillard 1981, p. 10). This consists not only of making the cartographic image and the topographic drawing take precedence over reality, but also in making people believe that they are derived from the latter. This is exactly what happens in an exemplary and paradigmatic way in the case of the state, which is transformed from a subject into an object. For Niklas Luhmann, there is no doubt that the modern concept of the state is a by-product of the printing press, especially the dissemination of maps and atlases, by virtue of which the state comes to be identified with its territory as it appears in its representations, whereas previously, in the absence of such representations, the notion referred to the bundle of rights at the disposal of the medieval lord (Luhmann 1989, pp. 88 and 109). But Luhmann does not explain how this semantic leap occurs; it is as if, in order to explain it, he is left with the effect of the "glance", which for Ptolemy was already the first vector of cartographic logic (see Figure 2.1).

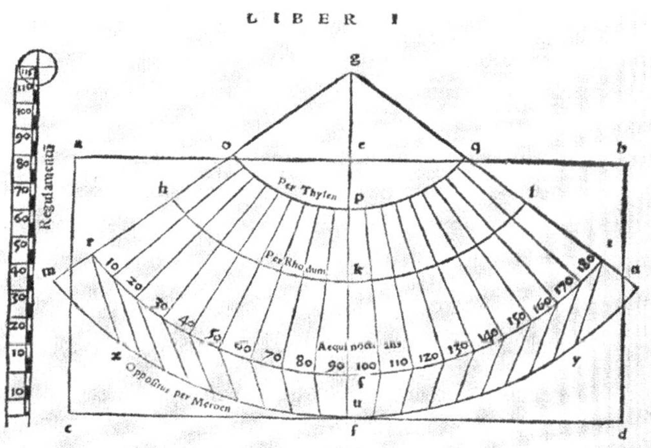

Figure 2.1. *The original source of modern space: Ptolemy's projection (source: Geographia Universalis Vetus et Nova, complectens Claudii Ptolemaei Alexandrini Enarrationis Libros VIII, apud Henricum Petrum, Basilae, 1540, p. 23)*

A valuable insight into the internal dynamics, the microphysics of the process in question can be found in the writings of Francesco Guicciardini. In 1512, the year the Medici family returned to power, Guicciardini sent a letter from Spain, where he was an ambassador, to his lord in Florence, in which he spoke of the benefits of the recent acquisition of Navarre by Ferdinand of Aragon. He first explained that "the things of Navarre" (i.e. its territory) "are quickly taken over at the foot of the Pyrenees mountains, by the band on that side" (Guicciardini 1825, p. 61). Today, we would say that the sovereign had seized the now French side of the Pyrenees, thus managing to prohibit the entry of the French into Spain, while opening up access to France for himself. Guicciardini considered that the annexation of Navarre was not so much valid from a fiscal point of view, since revenues of this type were marginal, but "for the sake of conformity" that it had with the other domains of which Ferdinand was already the lord of (Guicciardini 1825, pp. 61–62).

But what is meant here by "conformity"? So far, the answers have all pointed to the linguistic and cultural affinity of Navarre with the other domains of the peninsula. In his book on imperial Spain, Elliott (1963) acknowledges the vagueness of the term and relates it to its opposite, the *disformita* of customs and language referred to by Machiavelli in the third chapter of the *Prince*; the Florentine author refers to "mixed principalities" and recalls the happy inclusion in the kingdom of France of the "undeformed" provinces of Burgundy, Brittany, Gascony and Normandy. José A. Maravall (1991 [1972], pp. 108–109) also understands Guicciardini's "conformity" as the "uniformity" of language and customs, adding that the peninsular form of the Spanish land probably had a great influence on the determination of the modern state as a "closed space", but also, and at the same time, on the first state tendencies in Castile and Aragon. Both Elliott's and Maravall's responses take as an explanation what needs to be explained; they take for granted what was then a form of novelty, and therefore anything but normal. In fact, things are much simpler. But in order to account for this, we must first reverse the relationship we take for granted between geographical representation and reality, and begin to think that the latter is a copy of the former, and not the other way round as we are used to conceiving it. And since we are talking about the constitution of territory, that is to say, the area governed by the exercise of terror as explained in Justinian's code, we have to recognize that the map has the capacity to produce an essentially political force. Otherwise, we do not understand anything about modernity.

Before "conforming" to Aragon and Castile, Navarre was (and still is) contiguous with these two kingdoms, that is to say, situated in the mode of the first of the requirements that for Euclid define geometric extension: continuity. The same is true of the Île-de-France, the historical core of French unification, and of the regions that Machiavelli mentions. With the exception of Gascony, they are all in the immediate vicinity of the Île-de-France, progressively welded to it according to a plan of unification associating a mobile periphery to an immobile center. In more general terms, the “conformity” evoked by Guicciardini does not depend on any qualitative relationship (linguistic or cultural, for example) between one state and another; rather, it depends on a higher level relationship, on the common adherence of these states to the model of the cartographic structure, to the logic of the map and to its material constitution, whose continuity is the primary characteristic.

In fact, what gradually emerged from the 13th century onwards, in Europe and beyond, was what might be called the “mapping” of the face of the Earth itself, whose elements functioned like so many pieces welded together within an immense puzzle. Indeed, it is only on a map that contiguity becomes continuity. But this leads to another, even more problematic and far-reaching consequence: continuity is transformed into homogeneity of the fundamental cultural values that the subjects concerned share and which make them permeable to symbolic manipulation (Gellner 2006 [1983]). It is precisely in homage to this assumed principle of homogeneity that the contiguous expanse which, on some maps, remains white (i.e. empty), can be occupied, and thus materially annexed, by the very people who produced the map. Take, for example, the case of Muscovy and the progressive control, between the 16th and the end of the 19th century, of the immense plain stretching from the tributaries of the Danube to the Urals, and from the Russian forests to the Black Sea and the Caspian Sea. Materially, the Volga Plains were incorporated by a slow process of infiltration, but even earlier and following the ideas of the time they were incorporated in an immediate and cartographic form, as this was automatic and because of their direct contact with Muscovy (Sunderland 2004, pp. 17–18). The movement was thus guided by the logic of contiguity. This process is reminiscent of what Ludwig Wittgenstein, in the introduction to the *Tractatus* (Wittgenstein 1993 [1922]), assigned to the possible relationship between boundary and thought; but even before producing it “in language”, as Wittgenstein suggests, it is on the map that it is possible to “draw a boundary to thought”.

2.3. The map, order and modernity

The Tsar of Russia became Tsar of all the Russias and Emperor. This was in accordance with the same logic that underlies the entire modern state, represented in a programmatic way in the frontispiece of Thomas Hobbes’ *Leviathan* (1651) (see Figure 2.2), a formidable illustration of Ptolemaic projection: a process capable of reducing the world to a Euclidean geometric extension. As the engraving admirably shows, the gigantic body of the monster-like prince, *corpus civile* and *ecclesiasticum*, in which the citizens are allegorically gathered, rises to dominate the city and the countryside. His hands hold the symbols of the two powers: civil and religious. In the right hand is the sword, in the left the crosier. These two insignia have a convergent inclination toward an invisible point situated above the limits of the image, signifying that he governs from a distance. This point is called, in the language of cartographic technique, the point of projection, from which all maps, in a certain way, descend. It is precisely because of this functional determination that justice, which before Hobbes was something substantial, became a virtue, as the *bonum commune* (common good or Commonwealth in Hobbes' language) takes shape in society. It also becomes a jurisdiction, that is, a formal solution to the problem of coordinating clearly determined and determinable rights; a controllable act of assigning what each is entitled to. From justice, we thus pass to judicial authority, in the sense that what is good and justice are no longer thought of in terms of the community but depend on a meta-condition that proceeds in a formal, here expressly geometric, manner.

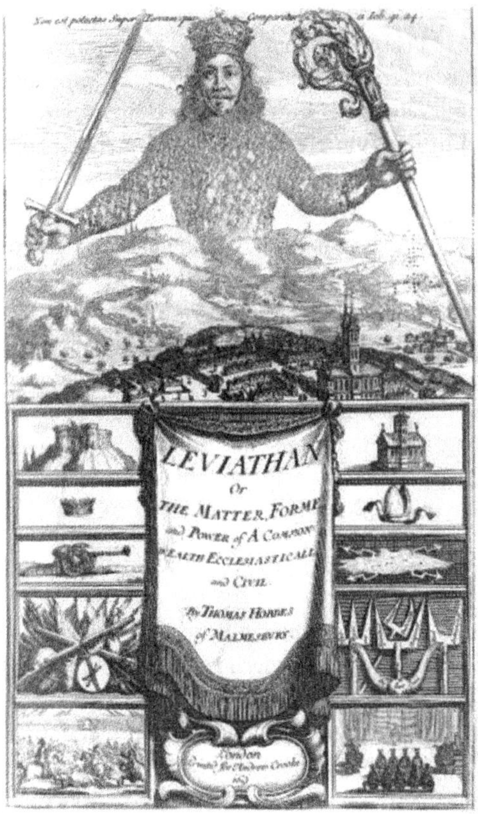

Figure 2.2. *Frontispiece to the first edition of Thomas Hobbes' Leviathan (1651) (source: Hobbes 2011 [1651])*

For Hobbes, the (geometric) order of the world is valid as a principle of justice; justice becomes the quality of the order itself and not, as before, of the sovereign. Previously – think of Plato's city – order was merely a reflection of justice. It is clear that for Hobbes, justice (the substance) is to jurisdiction (the form) what the city, understood as a set of relations between the inhabitants, is to the city reduced to its topographical skeleton. This happens not only because in both cases the passage from one to the other presupposes the expulsion of every concrete human being from the plane of determination, but also because this double determination is based on an identical representation, on a common matrix: that of the Euclidean geometric extension, whose properties are at the basis of the territory of every modern state, because they are at the basis of every cartographic image. In fact, the tremendous device of the upper part of the frontispiece is only valid as a signifier. The meaning, as for Saussure, lies in what is beneath the bar that cuts the image in two: we follow the logic of the table (as the map was called until the 19th century), which, as the authentic *machina machinarum* of modernity, feeds the whole logic by breaking down reality. As for the subjects of Leviathan, coherent and solidary elements of the same state body, not only can they not act autonomously, but, as mere anonymous bodies, they cannot even think with their own heads; the proof is precisely in the figure of the monster: only the head and the hands (the organs of thought and action) are made of a material that is not human, while the torso corresponds to the agglomeration of individuals. The head belongs by nature to the state and to it alone, in other words, to the map alone, of which the state is the image.

This figure thus expresses, among other things, the profound meaning of the Peace of Augsburg. Signed in 1555, it established a truce in the disastrous politico-religious conflict the then agitated Europe, and authorized each sovereign to impose their own religion on their subjects: *cuius regio eius religio*. Stephen Toulmin (1990) points out that such a principle made sense in the counties, duchies and small kingdoms of central Europe, as believers did not need to travel far to find a prince whose faith they shared, if they did not belong to a tolerant-free city where all denominations were admitted. However, in the case of large, long-unified kingdoms like France, where Protestants and Catholics lived side by side, the principle was not applicable; to avoid one side crushing the other, all that remained was to try to separate national loyalty from religious affiliation. This was the attempt of Henry of Navarre, who became Henry IV of France, and paid for it with his life in 1610. Toulmin points out that centuries later, it is difficult to understand why, for so long, the idea that a loyal French citizen could be either Catholic or Protestant was unacceptable.

The answer lies precisely in the link between physical continuity and cultural homogeneity, both of which depend on the topological qualities of the table, of which the mathematicians' plane and that of Euclidean geometry represent the idealized version. In the meantime, the problem can be formulated in terms of the relationship between Euclidean geometry and visualization. The latter would be endowed with two functions, one of image production, the other of norm production, and its merit would consist of transforming into a visual constraint what is in fact a logical constraint. In other words, the normative (i.e. regulating) function of the Euclidean visual *a priori* (here the tabular system and its idealization) produces the same effects "induced by logical inference by the conceptual elements of thought" (Reichenbach 1957, p. 42). Analogical inference consists of "transporting information from one domain (the medium) to another that lacks it, by a process of structural completion" (Markman and Moreau 2001, p. 372). Indeed, here the medium is the table; the objective is nationality, in which it is therefore possible to see a "fictitious ethnicity" (Balibar and Wallerstein 1991); analogical inference thus allows us to understand the ultimate, true meaning of what Guicciardini called "conformity" in relation to Navarre.

Thus, between the 16th and 17th centuries, Western thought became the vehicle of cartographic logic, assuming its nature. All modern understanding is thus based on a single movement, on an *adaequatio mentis et rei*: the adjustment between the mind, conceived as a flat table, and the thing, the face of the Earth, progressively reduced to the same pattern. This is a movement that has its distant origins in Aristotle's *Metaphysics* and *On the Soul*, in other words, after the revolution in which Clisthenes' Athens, through the use of geometric measurement, conceived the map as an operational territorial model (Lévêque and Vidal-Naquet 1964). In the modern era, in the wake of René Descartes' work, it is not only nature that is definitively transformed into "geometric space" and matter into "pure spatiality" (Borkenau 1984 [1934], pp. 347–350). The "long chains of reasons, all simple and easy, which geometers are wont to use to arrive at their most difficult demonstrations" (Descartes 1824 [1637], p. 19) become at the same time the model of all possible cognitive processes; they become the method: a term which literally means "that which comes after (*meta*) the concrete path (*odos*), and which together determines it, directs it", and which thus refers in all its aspects to the nature and function of maps. If we have never really been modern, as Latour (1991) suggests, it is not, contrary to what he concludes, because modernity rests on the distinction between nature and culture, but precisely for the opposite reason: because the idea of modernity consists above all in trying to make the knowledge of matter and that of the mind itself rise from the unique and coherent set of rules of the table, uniting the one and the other under the sway of the logic embodied in the Euclidean geometric framework. Descartes wrote in 1638 that nature is "one, simple, continuous, everywhere in agreement with itself", and that its knowledge should be "unique, simple, continuous, uninterrupted", so that "our contemplation of the whole and of the details is like a picture, or a mirror, representing very exactly the image of the Universe and of its singular parts" (Descartes 1987 [1638], p. 154) *Nomen omen*: the name is an omen.

The man who reflected on the so-called "book of the world" and the attributes of maps should have the term itself in his name – *Des Cartes*[2] – is the most amusing, but also the most significant and revealing detail in all of Western philosophy. Through Cartesian reduction, the Earth itself becomes literally (i.e. materially) a table. With the exception of three trips to France, from 1628 to 1649, Descartes resided in Holland, where the invention of watermills made it possible to transform ever larger expanses of sea into dry land, to convert what was liquid and mobile into what was solid and immobile. In just 1 year, 1640, for example, the arable land north of Amsterdam increased in area by about half: to remain wet meant poverty and idleness, while to be dry meant industry, freedom and wealth (Schama 1991 [1987], p. 25 and followings). The planning of the territories was guided by a simple idea of which the map was the model: what was fluid, amphibious and coalescing solidified.

2.4. The map, reason and rhetoric...

For Francis Bacon, a great enemy of "wet thought" (Bacon 2010 [1620]) truth is depicted as an island surrounded by the ocean, based on the recognition of the constant and mutual association between the form and nature of things. Only Hobbes, who was not afraid of water, took into account the fundamental plane that was found within the table and not individual bodies; in doing so, he identified the cartographic matrix with the process of legitimizing the state. For Hobbes too, as for the later scholastic thinkers and for Descartes, reason meant the capacity to construct syllogisms capable of leading to conclusions, by deduction from certain premises. This is an axiomatic procedure, the simple calculation by which (through additions and subtractions) we pass from general names to the denotation (which applies to us alone) or the attribution of meaning (in relation to others) of our thoughts (Hobbes 2011 [1651]). This is the reason why specialists in artificial intelligence today recognize him as their forefather (Dupuy 1994, p. 28). But at the same time, he systematically refers to the power of the visible, to the power of images, as a vehicle of ideal content. While Descartes returned several times to the function of "sensible images" in the representation of concepts (Rossi 1983, pp. 162–163), only Hobbes explicitly recognized them, albeit indirectly, as having the ability to instruct "in secret" and much more effectively than any precept.

2 The word "*cartes*" is French for "maps".

He had the opportunity to do so in the introduction to Thucydides' work on the Peloponnesian War, the English translation of which he published in 1628, in which he explained his conception of the relationship between history and philosophy. For Hobbes, Thucydides' greatness was not so much in the philosophical-moral interpretation of events, but in the clarity and vividness with which he places before the reader's eyes the consequences of the good and bad choices made by the protagonists of the conflict, in other words, in the insight and vividness of the narrative (Hobbes 1843 [1629], p. XXII). In his comparative analysis of history and philosophy with respect to mental impressions, Hobbes is in favor of the visual data of the former at the expense of the conceptual formulations of the latter. This prompts us to distinguish a first version of Hobbes from a second: the first is immersed in rhetorical thought ("wet" in the Baconian sense of the word) until the age of 40; the second shows himself increasingly fascinated by axiomatic calculation and its potential capacity to reach truly universal conclusions.

The *Leviathan* would be the union between these opposing forms of knowledge (Johnston 1986, pp. 22–24). But where and how does the reconciliation between reason and rhetoric occur, the latter being, according to Hobbes, the antithesis of rational discourse? The main difference between the two lies, for Hobbes, in the fact that the words with which we reason are always accompanied by a clear connection to the conceptions they represent, whereas, contrarily, rhetorical statements do not correspond exactly to thoughts. This means that while in rational discourse there is a two-way relationship between words and what they refer to, in rhetorical discourse this is not the case. The cartographic representation is exclusively composed of proper names (Bateson 1990 [1979]). On a map, the name signifies the thing and the thing is its meaning, as Wittgenstein intended. What Hobbes calls "rational" is therefore "cartographic", in the sense that it is the geometric nature of the cartographic structure of the representation that gives rise to the axiomatic character of reasoning.

However, at the same time, and precisely because it is rhetorical like all images, the cartographic representation silently conveys a surplus of meaning, irreducible to any rational expression or analysis. It is therefore this representation that ensures the co-presence, if not the conciliation, between rhetoric and reason. Indeed, to accompany his translation of Thucydides, Hobbes prepared a map of Greece, which he believed to be essential for understanding the text (see Figure 2.3). For Hobbes, as well as for Bacon, logic produces a "dry discourse", incapable of convincing the listener, as only images can. And the minds of his subjects, like the blank sheet of paper ready to receive the dictation of the public authority he evokes in *Leviathan* (Hobbes 2011 [1651]), expect nothing more than "a speaking image" through what is "unconscious or subliminal" (Johnston 1986, p. 25) on which political order depends.

Figure 2.3. *Ancient Greece, marking especially the places mentioned in Thucydides by Thomas Hobbes (source: The History of the Grecian War, faithfully translated from the original by Thomas Hobbes). For a color version of this figure see www.iste.co.uk/debarbieux/mapping.zip*

Until now, this order had been thought to be linguistic in origin, dependent on an image produced in the mind by words (Ball 1985, pp. 739–760). It was not understood that it could in fact only be a map. Just as it had not been understood, more generally, that it was only from the coincidence of tabular/mapping logic and the functioning of the world that modernity was born. Toulmin (1990) summarizes the philosophical program of modernity in three foundations: "certainty, formal rationality, and the desire to start with a clean slate". These apply as much to scientific analysis (e.g. the explanation of the movement of the planets or the tides) as to the production of the social and political world (e.g. the legitimacy of the nation-state system).

More specifically, the abandonment in the 17th century of humanist postures involved, according to Toulmin, a fourfold rejection:

– rejection of all oral knowledge in favor of written knowledge, so that formal logic replaced rhetoric;

– rejection of all particular knowledge in favor of universal knowledge, so that the study of the multitude of particular cases was replaced by that of general principles;

– rejection of all local knowledge in favor of general knowledge, so that the concrete diversity of phenomena and situations left room for the elaboration of abstract axioms;

– and finally, rejection of all temporal knowledge (i.e. punctual in relation to events and specific in relation to contexts), in favor of timeless knowledge, so that the transitory left the scene and the perennial took over.

In moving into the control room of knowledge, we must add to what Toulmin says that these rules erase almost all memory of their installation, thus all reflection on the nature of knowledge itself. Toulmin points out that the result is a series of presuppositions rather than hypotheses, because they do not reflect a documented trajectory as is usually the case with scientific hypotheses behind which postulates are hidden. Precisely because they are general in nature, these hypotheses have never been seriously studied; instead they have been unthinkingly accepted prior to their implementation. I am pointing out here only a few of these assumptions/postulates whose cartographic origin, admittedly less mediated and yet obvious, derives from a single movement, that of substituting the map for the nature of reality itself:

– the material substance of physical structure is essentially inert;

– a body cannot act where it is not;

– physical objects and processes cannot think or reason;

– God combined natural objects into stable systems; etc.

One could go on like this. Such postulates and others, devoid of logic and even of factual foundations, constitute the intellectual scaffolding that, from the end of the 17th century onwards, enabled Newton and his followers to proceed with the construction of modern physics. And Toulmin himself asks what was implicitly at stake beneath the surface of the texts of that time.

2.5. The map, constraint and self-organization...

To answer this question, it is necessary to reverse Reichenbach's classical approach mentioned above, and to think less in terms of logical constraint than in terms of visual constraint: in other words, reasoning is the product of perception, not the other way around. This reversal allows us to overcome the same opposition that Plato, for example in *The Republic*, establishes between the visible and the truth, by making the former dependent on the latter, thus reversing the relationship between the (cartographic) copy and reality (Earth). In defense of his own conception of "visual thinking", of the intelligence of perception, Rudolf Arnheim provides unwitting support for the thesis of the pre-existence, in cognitive terms, of the table in relation to everything else. According to Berkeley, he explains, perception (i.e. the mental image) can only refer to an individual instance, not to a general concept. It would therefore not be useful to think abstractly. But why then, Arnheim asks, in all fields are diagrams and drawings used "to think at a high level of abstraction?" (Arnheim 1987, p. 168). For Arnheim, mental image and schematic reduction coincide, and he cannot think of the former without the latter.

Yet, contrary to what Arnheim seems to claim here, there is no necessary coincidence between a pattern or drawing and a mental image. Even if one is not thinking of a singular person or object, but of something as large and composite as a piece of the earth's surface, one may well conceive of it in terms that are both general and alien to any diagrammatic schematization or reduction. For example, in terms of a pictorial landscape, not a map. But for Arnheim, mental image and diagrammatic reduction coincide; he cannot think of the former without the latter. This, in turn, presupposes the existence of the table, which thus presides over all cognition. Even in the case of a baroque painting, let us suppose that a table is naturally necessary. But in this case, it would not transmit to the image its own syntax (the one called here cartographic logic), as the schematic reduction suggests.

It is more or less universally felt that it is precisely in this transmission that all modernity resides, and it also fully affects the outcome of the bond between the state and the nation. Known terrorists have warned that things could have turned out quite differently in this respect, but that it is not yet healthy to explain precisely why they did exactly what they did. In addition to the nation-state, the modern era saw the emergence of at least two other forms of state in Europe, to some extent in competition with the former: on the one hand, the empire and, on the other hand, the politico-commercial network of transnational cities, such as those of the Hanseatic League (Braudel 1979). In their profound diversity, these two forms were opposed to the nation-state in at least two ways: for the explicit and declared cultural heterogeneity of their constituent peoples, and for the blurred, if not open-ended, conception of territorial boundaries which, in the case of the empire, potentially referred to those of the entire Earth. As Wallerstein (1980 [1974]) reminds us, the bourgeoisie could have actually, not just theoretically, become conscious of itself on a global scale, and pushed influential groups such as the international merchant bankers and many wealthy farmers in this direction. It was only with the fall of the Habsburg Empire, that is, during the last century, that the European bourgeoisies recognized the nation-state as the definitive sphere of their assertion, to be established according to the principle of the "nationalization of society": the fabrication of an artificial community born of the instrumental claim of a so-called unity of language and race. However, because the continuation of the territorial revolution began with the absolute monarchies (Balibar and Wallerstein 1991), it was necessary to clearly pinpoint the limits of the table (its edges, etc.) as the stable limits of its own world, so that the material form of the table could serve as a blueprint for the concrete, singular forms of it.

We are thinking, for example, of the tendency of the capital to be central, even in cities which are not foundation cities like Madrid or Ankara. This position can only be explained if it refers to the third and last of the properties that Euclid assigns to geometric extension, beyond continuity and homogeneity: isotropism, which requires all the parts that the table (or the state) is composed of, is functionally oriented in the same direction.

2.6. The map, production and society

Paraphrasing Pascal, Henri Lefebvre has written that the state has reasons that reason ignores, because they cannot be seen and since they are buried in laws, whose essential character is guided by issues of identity, formal universality and equivalence between groups and individuals (Lefebvre 1976). In fact, for legal historians, and contrary to what was true for the old regime, the legal and political order of the contemporary state presents no mystery: it is homologous to the structural principle of territoriality, which consists of the triad "unity, polarization and homogeneity" (Hespanha 1992, pp. 471–473). Now it is obvious that this triad is in turn the exact correspondent of the Euclidean trilogy mentioned so far, which remains the matrix of modern space: if homogeneity appears in each one, it is correct and legitimate to substitute continuity for unity, and isotropism for polarization. Here is the proof that cartographic representation has succeeded, in the modern era, in replacing the world in an operational mode, in all and for all.

According to a famous definition, political representation is based on an antinomy, as it means "to make visible and present to the mind an invisible being by means of a publicly present being" (Schmitt 1993 [1928], p. 347). Thus, in representation, "a higher species of being is concretely manifested" (Schmitt 1993 [1928], p. 347). This is the political version of a much more general rule, and the copy always tends to transcend the original. The question is to determine which is the copy and which is the original (Schwartz 1996, pp. 211–257).

Proportionalist theories, focusing on the descriptive aspect of representation (*Darstellung*: something that simply represents another) insist on similarity and exact correspondence, and it is the map that counts as the main metaphor of reference in this case (Fenichel Pitkin 1967, pp. 62–73). In the glorious year of the Revolution, Honoré-Gabriel Riqueti, Count de Mirabeau, addressing the states of Provence, exclaimed (Mirabeau 1834, p. 7):

> [S]ocial law [...] requires [...] that the different orders which come together as a body of the nation [...] enter into this whole only with the measure relative to their importance. The states are to the nation what a reduced map is to its physical extent: whether in part or in large, the copy must always have the same proportions as the original.

A few decades later, Swiss J.C. Bluntschli (1877, p. 60) would be even more precise:

> Just as the map represents mountains and valleys, lakes and rivers, forests and meadows, towns and villages, so the legislative body must in turn form a condensation of the parts that make up the people, as well as the people as a whole, according to real relationships [...]. The value of each part is determined by its significance as a whole and for the whole. The relationships are organic, the scale is national.

These two excerpts show the transition from the medieval organicism of the "republic" of the body, which is still evident, to a function of proportionality which, with the presence of scale, is unequivocally dependent on modern spatial abstraction (i.e. cartography). And it is no coincidence that these extracts coincide with the end of the great cartographic works that ensure the geodesic portrait of the countries of their respective authors: we think of the great Cassini Map of France, whose last survey took place a few days before the storming of the Bastille (Gallois 1909), and in Switzerland the Dufour Map (Marquis 1987), whose 1:25,000 scale version appeared in 1873. In between, in Prussia on the eve of the 1848 revolution, knowledge of the topographic map ceased to be the preserve of the military; it became public, acquiring a general and widespread character, in order to meet the needs of civil construction of the territory (Stavenhagen 1900).

Cartographic representation is even more indispensable when it entails accounting for the other value of the term "representation"; no longer in the proportional sense, but for what relates to representation by mandate or delegation (*Vertretung*: something that acts, i.e. decides, in the place of another) within parliamentary democracies.

In the accumulation of linguistic uses of the term "representative-representation" (*Repräsentation*), over time, Hasso Hofmann has emphasized the one related to the representation of a totality understood as the presentification, personification or crystallization of a multitude of people in the actions of one of its parts, identified with certain criteria (Hofmann 2007 [1974]). The first problem that the French Estates General inaugurated on May 4, 1789 had to solve was this choice of criterion. The revolutionary solution would be the transformation of the Estates General into a National Assembly. For Starobinski (1973), in recommending that the three states deliberate separately, Louis XVI, in the early days of the revolution, had acted in perfect coherence with the feudal spirit, which applied its zonal worldview to the social world. The Third Estate was guided instead by the ideal of a "homogeneous totality", following the model of the "homogeneous and isotropic space" of celestial mechanics, permeable in all directions to the force of universal gravitation. But in his analysis, Starobinski does not take the ultimate, yet crucial step; he does not realize that before being celestial or terrestrial, the space of the Third State is the space of cartographic representation, which the space of the sky and that of the Earth have depended on in the order of ideas for centuries. Let us reread the fateful definition (Sieyès 1789, pp. 112–113):

> I see the Law at the center of an immense globe; all citizens, without exception, stand at the same distance on the circumference, and occupy only equal places; all depend equally on the Law, all offer it their liberty and their property to be protected; and this is what I call the common rights of the citizens, by which they all resemble each other.

In this passage, there is more than the atomistic postulate or the idea of the individual as a numerical unit that has been identified at the heart of the contradiction of Jacobin discourse (Jaume 1989, pp. 151–191). The equidistance of citizens before the law is only possible if the globe to which Sieyès refers is in fact a circle[3], according to the coincidence established at the time of the Renaissance between this one and the sphere (Brendel 1977). This is to say that if Sieyès

3 For if the globe were really a globe, the citizens would not be on the same plane at all, and therefore, depending on their position, they would be at very different distances from the center.

expresses himself in terms of spheres, he thinks according to the logic of the map, and according to the structural and systematic correspondence between individuals and the geometric points of a plane, so that the equality between the former is exemplary over that which applies to the latter. And it is precisely the unthinking (or taken-for-granted?) insertion of the cartographic model into political discourse (most evident in the case of the constituent to whom we owe the most profound and powerful formulation of representative government) that is at the root of the equality in question. Thus, the distinction proposed by Lefebvre (1974) between spatial practice (perceived space), representation of space (thought space) and representation space (lived space), which still inspires English-language rhetoric in the field of urban and territorial studies, short-circuits itself and proves to be useless, as no distinction between the three spaces is possible. This distinction concerns three articulations of the same model, the cartographic model, whose progressive colonization on the conception and practice of the world they reflect (Lefebvre 2000 [1974]).

Abbé Sieyès had no doubt: the territory is a blank map, to be divided, articulated and qualified according to the needs of a purely subjectivist rationality (Mannoni 1994, pp. 274–275; Mannori 2008, pp. 23–56), in other words, a reason embodied in Euclidean geometry that presupposes a cartographic order of relationships between subject and object. What Lefebvre calls the "social production of space" does not exist. Only the spatial (i.e. cartographic) production of society exists.

2.7. References

Arnheim, R. (1987). *Intuizione e intelletto: nuovi saggi di psicologia dell'arte*. Feltrinelli, Milan.

Bacon, F. (2010). *The Great Instauration and the Novum Organum*. Kessinger Publishing, Whitefish [Original published 1620].

Balibar, E. and Wallerstein, I.M. (1991). *Race, Nation, Class: Ambiguous Identities*. Verso Books, London.

Ball, T. (1985). Hobbes' linguistic turn. *Polity*, 17(4), 739–760.

Baudrillard, J. (1981). *Simulacres et simulation*. Galilée, Paris.

Bluntschli, J.C. (1877). *Théorie générale de l'État*. Guillaumin, Paris.

Borkenau, F. (1984). *Le passage de la vision féodale à la vision bourgeoise du monde*. F. Alcan, Paris [Original published 1934].

Braudel, F. (1979). *Civilisation matérielle, économie et capitalisme : XVe–XVIIIe siècle III : Le temps du monde*. Armand Colin, Paris.

Brendel, O.J. (1977). *Symbolism of the Sphere: A Contribution to the History of Earlier Greek Philosophy; with 30 Plates*. Brill [Online]. Available at: https://brill.com/view/title/5708 [Accessed 20 October 2020].

Chabod, F. (1961). *L'idea di nazione*. Laterza, Bari.

Descartes, R. (1824). *Discours de la méthode pour bien conduire sa raison et chercher la vérité dans les sciences*. Levrault, Paris [Original published 1637].

Descartes, R. (1987). *Œuvres philosophiques, volume 2*. Garnier, Paris.

Dupuy, J.-P. (1994). *Aux origines des sciences cognitives*. La Découverte, Paris.

Elliott, J.H. (1963). *Imperial Spain: 1469–1716*. Cambridge University Press, Cambridge.

Farinelli, F. (2003). *Geografia. Un'introduzione ai modelli del mondo*. Einaudi, Turin.

Farinelli, F. (2009). *De la Raison cartographique*. CTHS, Paris.

Fenichel Pitkin, H. (1967). *The Concept of Representation*. University of California Press, Berkeley.

Gallois, L. (1909). L'Académie des Sciences et les origines de la carte de Cassini. *Annales de géographie*, 18(99), 193–204.

Gellner, E. (2006). *Nations and Nationalisms*. Blackwell Publishing Ltd, Oxford [Original published 1983].

Guicciardini, F. (1825). *Legazione di Spagna*. Presso Niccolò Capurro, Pisa.

Hauber, E.D. (1727). *Nützlicher Discours von dem gegenwärtigen Zustand der Geographie besonders in Deutschland*. Wagner Ulm, Ulm.

Hering, G.C. (1728). *Gedancken von der Nützbarkeit und Nothwendigkeit der Geographie, auch von der Methode und Ordnung dieselbe zu tractiren, über des Hrn. Doct. und Prof. Leysers von der allgemeinen Art darin abgehende Meinung, mit dazu dienlichen Anmerckungen entworfen*. Rüdiger, Berlin.

Hespanha, A.M. (1992). L'espace politique dans l'ancien régime. *Boletim da Faculdade de Direito: Universidade de Coimbra*, 58, 455–510.

Hobbes, T. (1843). *The English Works of Thomas Hobbes of Malmesbury, Volume VIII: The History of the Grecian War by Thucydides*. John Bohn, London.

Hobbes, T. (2011). *Leviathan*. Pacific Publishing Studio, Seattle [Original published 1651].

Hofmann, H. (2007). *Rappresentanza-Rappresentazione. Parola e concetto dall'antichità all'Ottocento*. Giuffrè, Milan [Original published 1974].

Jaume, L. (1989). *Le discours jacobin et la démocratie*. Fayard, Paris.

Johnston, D. (1986). *The Rhetoric of Leviathan. Thomas Hobbes and the Politics of Cultural Transformation.* Princeton University Press, Princeton.

Kant, I. (1998). *Critique of Pure Reason* (translated by P. Guyer and A.W. Wood). Cambridge University Press, Cambridge [Original published 1781].

Koselleck, R. (1979). *Le Règne de la critique.* Minuit, Paris [Original published 1959].

Latour, B. (1991). *Nous n'avons jamais été modernes. Essai d'anthropologie symétrique.* La Découverte, Paris.

Lefebvre, H. (1976). *De l'État, volume 1.* Union générale d'éditions, Paris.

Lefebvre, H. (2000). *La production de l'espace.* Anthropos, Paris [Original published 1974].

Lévêque, P. and Vidal-Naquet, P. (1964). *Clisthène l'Athénien. Essai sur la représentation de l'espace et du temps dans la pensée politique grecque de la fin du VIe siècle à la mort de Platon.* Macula, Paris.

Leyser, P. (1726). *Commentatio de vero Geographiae Metrodome.* Typis Pauli Dieterici Schnorrii, Helmstadii.

Luhmann, N. (1989). *Gesellschaft und Semantik. Studien zur Wissenssoziologie der modernen Gesellschaft, B and 3.* Suhrkamp, Frankfurt.

Mannoni, S. (1994). *Une et indivisible. Storia dell'accentramento amministrativo in Francia. Volume I.* Giuffrè, Milan.

Mannori, L. (2008). La nozione di territorio fra antico e nuovo regime. Qualche appunto per uno studio sui modelli tipologici. In *Organizzazione del potere e territorio. Contributi per una lettura storica della spazialità*, L. Blanco (ed.). Angeli, Milan.

Maravall, J.A. (1972). *Estado moderno y mentalidad social (Siglos XV a XVIII). Volume 2.* Revista de Occidente, Madrid.

Markman, A.B. and Moreau, C.P. (2001). Analogy and analogical comparison in choice. In *The Analogical Mind. Perspectives from Cognitive Science*, Gentner, D., Holyoak, K.J., Kikinov, B.N (eds). MIT Press, Cambridge, London.

Marquis, J.M. (ed.) (1987). *G.H. Dufour, l'homme, l'œuvre, la légende. Le Portrait Topographique de la Suisse.* Maison Tavel, Geneva.

Mirabeau, H.G.R. (1834). *Oeuvres de Mirabeau. Volume 1 : Discours prononcés à la tribune nationale.* Chez les libraires-editeurs, Paris [Original published 1789].

Reichenbach, H. (1957). *The Philosophy of Space and Time.* Dover, New York.

Rossi, P. (1983). *Clavis Universalis. Arte della memoria e logica combinatoria da Lullo a Leibniz.* Il Mulino, Bologna.

Schama, S. (1991). *L'Embarras des richesses.* Gallimard, Paris [Original published 1987].

Schmitt, C. (1993). *Théorie de la constitution*. PUF, Paris [Original published 1928].

Schmitt, C. (2001). *Le nomos de la Terre*. PUF, Paris [Original published 1950].

Schwartz, H. (1996). *The Culture of the Copy: Striking Likeness, Unreasonable Facsimiles*. Zone Books, New York.

Sieyès, E.J. (1875). Qu'est-ce que le tiers état ? [Online]. Available at: https://fr.wikisource.org/wiki/Qu%E2%80%99est-ce_que_le_tiers_%C3%A9tat_%3F [Accessed 18 November 2019].

Starobinski, J. (1973). *1789 : Les emblèmes de la raison*. Flammarion, Paris.

Stavenhagen, W. (1900). Die geschichtliche Entwickelung des preussischen Militär-Kartenwesens. *Geographische Zeitschrift*, 6(H.2), 435–480.

Sunderland, W. (2004), *Taming the Wild Field. Colonization and Empire on the Russian Steppe*. Cornell University Press, Ithaca.

Toulmin, S. (1990). *Cosmopolis: The Hidden Agenda of Modernity*. University of Chicago Press, Chicago.

Wallerstein, I. (1980). *Le système du monde du XVe siècle à nos jours*. Flammarion, Paris.

Wisotzki, E. (1897). *Zeitströmungen in der Geographie*. Duncker und Humblot, Leipzig.

Wittgenstein, L. (1993). *Tractatus Logico-Philosophicus*. Gallimard, Paris [Original published 1922].

3

Farewell to Maps: Reformulating Critical Cartography in the Digital Age

Pierre GAUTREAU[1] and Matthieu NOUCHER[2]

[1] *UMR PRODIG, University of Paris 1 Panthéon-Sorbonne, France*

[2] *CNRS, UMR Passages, University of Bordeaux, France*

3.1. Introduction

For decades, representations of space have been the subject of analyses that decipher their construction and their sociospatial effects from a political point of view. *Critical cartography*, a research trend that emerged in the 1970s and 1980s, was thus decisive in the adoption of a denaturalizing approach to maps. Today, however, the changes brought about by the digital transition have not yet been sufficiently integrated into this analytical framework. The time when maps were only seen as mediators of relations of domination seems to be over, in favor of diversified approaches that question the complex interactions between cartographic production, the forming of territories, processes of spatial visibilization/invisibilization, etc. The thesis of this chapter, however, is that the changes brought about by digital technology have disrupted the objects and questions of critical cartography to such an extent that the concepts, methods and objects to which attention is given must evolve in unison. Although, as early as the 1990s, many researchers were quick to perceive that the world they were studying was changing, political analyses of the new digital issues have remained scattered and fragmentary. Our purpose here is therefore manifold: to draw up a reasoned assessment of the new conceptual and methodological challenges of digital technology for critical

The Politics of Mapping,
coordinated by Bernard DEBARBIEUX and Irène HIRT. © ISTE Ltd 2022.

cartography (first section), to point out some of the historical shortcomings of the discipline that have made the analysis of the new phenomena more difficult (second section) and finally, to propose methodological *aggiornamento* avenues that can better account for the new ways in which power and representations of space interact today (third section). In short, we propose to reformulate critical cartography in the digital age.

3.2. Farewell to maps: when the digital world disrupts the objects and questions of a research current

Even before posing new questions for critical cartography, which we will give at the end of the chapter, the digital world disrupts the questions that could be described as "conventional" within this movement.

3.2.1. *The main issues of conventional critical mapping*

Critical cartography has retained from its beginnings the simple idea that maps serve power. In a general context of questioning the objectivity of science (Jasanoff 2004), the founding work of British geographer and map historian John Brian Harley truly launched the critical cartography movement in the 1970s and 1980s (Harvey 1989). By distancing himself from functionalist approaches and emphasizing the cognitive scope of maps, he invited the social sciences to rethink them as forms of socially constructed, subjective and ideological knowledge. In France, it was undoubtedly Mark Monmonier's *How to Lie with Maps* (1991), which contributed the most to popularizing this first stage of critical cartography. In the 1980s and 1990s, a historian movement strengthened around the study of the map as an instrument for the formation of the modern State (Kain and Baigent 1992; Wood 1992; Touzery 2007; see also Chapter 5 of this book).

To answer the seminal question "How can maps be used to exercise power?", critical cartography has gradually brought considerable nuance. As Pickles (2004) suggests, the Foucauldian and post-modern turning point of the 1990s complicated the vision of cartography as a mere "repressive" or authoritarian exercise of power, in favor of an approach in which the map became part of a "power-knowledge". He argues that "mapping as a power-knowledge also functions as productive power with to constitute objects, identities and practices that are part of (and constitutive of) our world" (Pickles 2004, p. 114). More subtle forms of domination and control, circulating, internalized by individuals and no longer imposed from the outside by force, are highlighted.

This second stage of critical cartography thus opens up two research programs, which are similar but not identical: on the one hand, it is a question of understanding cartography as one of the techniques for the territorialization of power, in an explicitly Foucauldian vein; on the other hand, it is to analyze it as a vector of dissemination of hegemonic visions within the population, which could be described as "acculturation". Without clearly distinguishing between the two, Pickles shows this dual agenda of post-modern critical cartography in his 2004 book. He argues that maps and cartographic institutions are to be seen as "archetypes of power-knowledge [...] that is, a discourse, practice and set of institutions that delimit potentialities through the control of space-time-action, and thereby produce certain types of subjects, actors, and places" (Pickles 2004, p. 111). The idea of territorialization is fundamentally geographical, but we should note that it is more often people from countries on the "winning" side of history who write these essays. In 1988, Tongchai Winichakul defended his thesis on the role of cartography in the formation of the State that would become Thailand, *Siam Mapped: A History of the Geo-Body of a Nation*. His central idea is that the cartographic enterprise shaped the Siamese territory and nation, and that "maps operated as technologies of territoriality" (Pickles 2004, p. 117). In short, the map precedes the territory[1].

The idea of acculturation, on the other hand, is approached in many ways and by various disciplines, for instance, take France; we can mention the research of Serge Gruzinski on the acculturating effects of the circulation of Spanish maps among the Nahuatl elites, who were incorporated into the Spanish Empire during the 16th century (Gruzinski 1987, 1993 [1988]). Even if this defeated nobility was able to mobilize maps made according to Spanish norms to exercise a form of resistance against their new masters, the simple fact of having recourse to the representations of these "Others" contributed to profound changes in their culture and social organization. In the same vein, the controversy surrounding the so-called "Peters projection" (Crampton 1994) had already initiated this questioning of acculturation. In 1974, this historian's proposal to discard the Mercator projection due to its "racist" character (Pickles 2004) was already clearly part of this trend, carried in this case by post-colonial activism: it was a question of denouncing this widely used projection, which artificially minimizes the surface area of the lands colonized by

1 What would become the antiphon of Foucauldian-inspired critical cartography found one of its emblems in Winichakul's work. In 1991, Benedict Anderson explained in the preface to the second edition of his famous book *Imagined Communities* (Anderson 1991) that it was his discovery of Winichakul's work that led him to incorporate on the map in Chapter 10 of his book.

Europe, as a means of acculturating the populations of the "Third World" to the representation of the world of their oppressors[2].

3.2.2. *What the digital world does to critical mapping: Russian doll-like difficulties*

The digital transition that has been accelerating since the late 1990s has brought about a number of interrelated difficulties that complicate the analysis of map production and representations of space from a political perspective. The first difficulty is the material change in the cartographic objects themselves, which shatters the map into multiple informational artifacts (from virtual globes to GPS tracks, from geographic databases to geolocation applications, etc.): it is therefore now difficult for critical currents to clearly establish *what they are talking about.* The second difficulty is the circulation of these objects. Promoted by the Internet, these are more difficult to follow than they were in the past, since they are made up of multiple digital objects that can aggregate or disintegrate *ad infinitum*: under these conditions, identifying *who does what* in the contemporary mapping process poses formidable methodological questions for anyone interested in power relations, for which it is necessary at the very least to identify the various stakeholders. As a consequence of these first two difficulties, it is nowadays difficult to grasp precisely which dimensions of power are affected by these new forms of cartographic production: if the objects are difficult to grasp, and the actors multiple and vague, how can we succeed in establishing the way in which power is instituted and circulates?

3.2.2.1. *Explosion of objects and categories*

The persistent focus of critical cartography on the *map* object makes it difficult to understand recent developments, and in our view the very term *cartography* brings about more problems than solutions for relevant policy analyses. Indeed, everything seems to indicate that we are no longer systematically dealing with maps in the strict sense of the word in the context of the digital transition, but with more complex and shifting objects. Virtual globes, for example, such as *Google Earth*, are above all puzzles made of images. First, critical cartographers only began to study these new products, which were initially considered uninteresting because they were

2 This example is based on work that, prior to the formal advent of a field of knowledge clearly labelled as "critical cartography", was already practicing similar approaches. For example, there are many expressions that could be classified as critical cartography, but which predate the work of Brian J. Harley. This is what Crampton and Kryger (2006) point out. We refer the French-speaking reader who wishes to learn more to the third chapter of Gilles Palsky's (2003) dissertation.

produced outside the sphere of patent experts and intended for the "general public", at a very late stage. Second, they did not say openly enough that their practice was gradually shifting toward critical approaches to a multitude of representations of space, including the map, which was no longer exclusive. Nor did they sufficiently explain what this implied in terms of methodological *aggiornamento*: traditionally, the political criticism of maps identifies an author or a person responsible for a cartographic production, questions their intentions and representations, and analyzes the way in which their maps carry these and how they will interact with other actors. However, when maps consisted of geographic information systems (GIS), from digital *layers* produced by different actors who do not know each other and who have worked separately with different objectives, it is understandable that the analysis of the representations and intentions of which the maps are the bearers is made more complex: each layer may implicitly "embody" a specific representation, which will have political effects, regardless of the intentions of the person who makes a map with it. A simple example of this situation: it is now commonplace to create maps of tropical countries using forest vegetation layers produced at the continental scale by international agencies or nature conservation NGOs (on this topic, see also Chapter 6 of this book). These vegetation layers are not a neutral representation of forests, but the product of technical choices with strong political consequences. For example, classifying agroforests as *primary forests*, and therefore requiring specific protection, may lead some governments to drive rural populations that depend on this environment out of their territories. However, as long as this type of layer is included in a map without being subject to specific criticism, the final map *embeds* or *inherits* implicit political biases.

The digitization of spatial data thus complicates, if not makes impossible, the constitution of clear corpora according to the conventional categories of policy analysis. As has recently been pointed out, digital data have become *malleable*, in the sense that they are easily modified by people other than those who created them (Joliveau *et al.* 2013). Associated with this *malleability*, their reproducibility, which has become extremely easy at generally marginal costs, explains why data can have many lives outside their original media. The founding categories of research in information and communication, that of the *producer* and the *user*, are thus potentially turned upside down. A growing number of digital data users will modify them so that they better correspond to the uses they want to make of them, combining them with other data sets, and in turn becoming producers of these data (hence the term "*produser*"). Digital technology has thus blurred the boundary between the notion of the *production* and that of the *circulation* of information. Although in the past information was generally produced within a stabilized framework and then circulated without being altered too much, today this framework explodes and circulation becomes one of the ways in which information is produced.

Another effect of the digital transition is the increasing porosity between two previously relatively sealed worlds: the public and the private. It is increasingly common for data sets to associate several producers of different status (Joliveau *et al.* 2013). The democratization of individual positioning tools (GPS) has also changed the situation by multiplying the number of contributors to spatial databases in certain fields (amateur naturalism, road applications, etc.). The large national and global databases are therefore partly made up of data collected by individuals not affiliated with a public institution, which gives them a different status from conventional databases, made up solely of data collected by specialists. Some of these databases, such as Geonames[3] for toponymy or OpenStreetMap[4] for cartography, are no longer administered by public entities but by groups that are keen to make these digital commons live beyond institutional frameworks.

Another example is interoperability, a computer function that is increasingly present in spatial database management systems. It allows, from one interface, one to consult, view and often download data that is located on other servers and created by other institutions. This being the result of long research and negotiation to define common standards, interoperability has, in the last 10 years, become a reality that is fully integrated into the daily life of cartographers and other geomatics specialists. It allows, for example, web platforms to *harvest* remote databases, and thus to aggregate data from different sources. This function therefore questions the outlines of a spatial database, since these can vary according to the requests made by the user at the *harvesting* interface. Finally, when data are shared on the Internet, when they are open, the identity of the receiver is questioned, potentially extended to any person equipped with a means of access. The fact that people can take hold of this information even though they were not targeted by its formatting profoundly changes the analyses of its reception. This new "regime of digital truth", analyzed by Rouvroy and Stiegler (2015), has hardly been worked on in the field of critical cartography.

3.2.2.2. *Difficulties in understanding the effects of territorialization and acculturation*

From the previous section, we can see that it is difficult to clearly identify "who influences whom" when the objects that mediate power relations (maps, spatial data layers, etc.) are difficult to grasp. This indeterminacy also makes it very difficult to

---

3 http://www.geonames.org/.

4 https://www.openstreetmap.org/.

clearly identify how the processes of territorialization and acculturation take place. Take, for example, the question of cartographic sovereignty, that is, the conditions that allow an actor to be autonomous in the representation and qualification of their territory. Since at least the 19th century, this sovereignty has been a major element in the construction of national identities, under the leadership of the State, which disseminates its official maps in the public space. The conjunction of satellite technologies, digital technology and geopolitical relations means that it is now possible to map foreign territories and resources from a distance, while most territories are now "photographed" and mapped by others. Understanding how these cross- and distance-readings influence the geopolitical game and national representations of territories and their issues is a puzzle for critical cartography. As has been shown for environmental issues (Gautreau 2021), politically analyzing these processes would require monitoring the software used to measure the environment, the logics of their dissemination, the role of commercial wars between the companies that provide them, the strategies for training bodies of geomatic experts by using international cooperation agencies, etc. The diversification of actors capable of producing environmental information at a distance increases this imbroglio a little more. The major emerging countries are setting up space programs, with varying degrees of success in their plans to equip themselves with their own Earth observation satellites, while large private companies are also setting up their own environmental mapping programs[5].

This deregulation from above (by actors operating at the national or even supra-national level) is accompanied by deregulation from below, carried out in a diffuse manner by actors with diverse profiles, acting at the local level, and who mobilize geographic information to turn the power of maps upside down and seize them to make their demands. This counter-mapping is not new, but such initiatives – led by environmental activists, Indigenous communities or even groups of committed artists – tend to multiply by taking advantage of the opportunities offered by digital technology. In challenging the cartographic authority of States, this double deregulation opens up an era of post-sovereignty cartography (Noucher 2020) that can appear both as a symptom and as an agent of the crisis of nation-States confronted with globalization (Desbois 2015).

Given this context, it is clear that we must reformulate in new terms the old question of *resistance*, which was made explicit in the 1990s by critical cartography: what does it mean today to resist legibility by others, when these neo-cartographers are elusive hydras or are thousands of miles apart, never go into the field, and disseminate their analyses on the Web without our having any right of review?

5 See Amazon and its project: https://www.globalforestwatch.org/.

3.2.3. *Methods maladapted to digital technology changes*

In the mid-1990s, however, critical approaches found a new resonance, but without significantly renewing this research current. Some of the researchers working on GIS and developing geomatics came together in the United States under the name of *GIScience*, and founded a research current around *Critical GIS*, which questioned the social implications of these technologies. Toward the end of the 1990s, critical GIS became a sub-discipline of digital geography (Desbois 2015), which saw the development of different branches around, for example, participatory GIS intended to introduce geomatic techniques into participatory democracy actions (Rambaldi 2005, pp. 538–543). However, it can be argued that many of the important issues identified by this stream, such as that of the complexification of data processing chains described by Elwood (2009), have not been translated into extensive investigations of the subject. More worryingly, critical approaches to GIS still remain in the extreme minority.

Faced with these changes, critical cartography progresses by methodological trial and error: the urgency is first to succeed in identifying the new objects that are emerging. The difficulty in defining what is emerging is then compensated for by the invention of vague notions, to which an attempt is made to give more rigorous content *a posteriori* (GIS/2, neogeography, geoweb, etc.). While the speed of technological developments does not help researchers to think more calmly about the world that we live in, it can also be said that the theoretical weight of the early days persists and limits new approaches. Much of the promise (and much of the fear) of the new digital manifestations of the *map* is deeply rooted in the original paradigm (map = dominant power) – these approaches perceive ongoing processes to be less likely to lead us into a new paradigm; rather, they reinforce previous ones. Jeremy Crampton's approach is emblematic of these difficulties in rethinking cartography in the digital age in a way that is off the path that was traced during the 1980s and 1990s. When confronted with the arcana of cyberspace, he reuses Foucauldian categories in a truncated sense, with power manifesting in a one-way fashion: surveillance of citizens, mapping of "crime", governmentality, etc. (Crampton 2003). We can probably speak of path dependency here, when it comes to this emblematic author of critical cartography, rather than questioning the relevance of these concepts. He seems to be looking in the abundance of digital innovations for examples that fit a well-established conceptual scheme.

This difficulty in thinking about the new things that are happening can also be explained by the lack of interest among geographers in tools for the *general public*, such as virtual globes, whose technical operating logic escapes them. According to Michael Goodchild, this lack of interest leads to a marginalization of critical approaches. He notes that there are proportionally far fewer researchers engaged in this path today than in the mid-1990s: "Critical GIScience is in danger of becoming almost invisible" (Goodchild 2014, p. 5). The new cartographic promises associated with big data seem, however, to be marked by a resurgence of positivism or a form of "digital neo-positivism" (Mosco 2013) and require more than ever to be deciphered from a political angle.

3.3. The shortcomings of critical mapping exacerbated by digital technology

While critical cartography has not been without its debates, we believe that these are exacerbated by the digital transition. In this section, we highlight three methodological problems that, since the early days of critical cartography, have caused it either to overestimate certain political effects of maps or, conversely, to lose sight of more subtle processes of power. While one might have expected that digital changes would highlight these problems on their own, this is not yet the case.

3.3.1. *The "Harlesian" of reflections on the finished product*

While Harley's work rightly contributed to a "broadening of the horizon of cartography" (Gould and Bailly 1995, authors' translation), which led to the denaturalization of the map by conceiving it as a *text* whose meaning had to be unraveled, it also installed for too long a critical method less interested in the process of map construction than in the analysis of the map as a finished product. Here, we can see the gulf that separates "Harlesian" cartography from the sociology of science that established its methods at the same time. The former fixes its attention exclusively on a moment, a fixed expression (the map) of a dynamic process of representation of space (mapping). The latter, on the other hand, pays full attention to the process itself and interprets the results of this process as often precarious forms of the "stabilization" of reality, always subject to reinterpretation[6]. This prioritization of the *product* over the *process* was denounced very early on within critical cartography itself, particularly by the practitioners whose fields differed greatly from Harley's Britain. For example, as early as 1991, Robert Rundstrom denounced the fact that critical cartography at the time unacceptably restricted the kinds of maps and mapping acts it chose as objects of study to ones of Western cultures, and argued for the inclusion of Indigenous cartographies where the process may be more significant than textuality (Rundstrom 1991).

6 This is particularly the case with work that has mobilized the actor-network theory for over 20 years, from Harvey and Chrisman (2004, pp. 65–80) to Trouillet (2018).

This fixation on the "map-object" is rendered partially obsolete by a multitude of processes that now affect the production of representations of space (discussed in the previous section), making them evolving and subject to multiple modifications in the course of their existence, often impossible to observe as a "finished product". On the cartographic interfaces of the geoweb, one could consider that there are as many maps as there are clicks! However, despite the fact that the focus on the map as a finished product is no longer possible exclusively in view of the concrete developments in contemporary cartographic production, it still persists in many works. This is notably what Matthew Edney argues in a recent book, *Cartography: The Ideal and Its History* (2019). In it, he argues that cartographic research is subtly sidetracked by a set of preconceptions that form a kind of ideal. This "cartographic ideal" (Edney 2019) would, he argues, be a pervasive and masked cultural belief system that maintains that there is an unequivocal category of objects called "maps", all of the same type. Whatever their form and function, these objects can be described with the same vocabulary and can be evaluated according to the same standards. The diversity of maps is reduced to a single function, that of representing spatial realities. Anything that deviates from this is removed from the official designation of a "map". Edney then demonstrates that even the critical cartography he calls "sociocultural criticism", while moving in the right direction, remains a victim of some of the preconceptions associated with the cartographic ideal and misses certain objects (close to but different from the conventional map) as well as certain processes, undetected because they are not foreseen by the theoretical framework. From Edney's perspective, we believe that the pervasiveness of this "cartographic ideal" largely explains the blindness of many analyses to cartographic processes related to the radical novelties brought about by the digital world.

3.3.2. *The taste for successful maps*

Another continuing trend in critical cartography is a strong interest in successful devices, that is, those that enable those who master them to increase or maintain their power. Critical cartography is thus often the historian of victorious representations that have contributed to strengthening of the already powerful. This focus can no doubt be partly attributed to the logical disappearance of failed cartographic projects, both from memory and from archives, according to the adage that geography is rarely written by the "losers". More seriously, it is also possible to denounce the deliberate selection of cases that corroborate their theory with critical cartographers who want to demonstrate the effects of domination that their object generates (Palsky 2002). In doing so, this literature obscures an essential part of the career of cartographic enterprises, which is marked by failures, social resistance to their deployment and difficulties in achieving their objectives. In this respect, it is important to re-read James Scott's now classic book, *Seeing Like a State* (1998), too often presented as a chronicle of the unambiguous shaping and normalization of society by the modern State. His first chapter, "State projects of legibility and simplification", rightly emphasizes the many obstacles to this effort of standardization: Prussian scientific forestry in the second half of the 20th-century simplified forest ecosystems to such an extent that it quickly led to a reduction in wood production following the spread of diseases and the decline of plantations planted on unsuitable soil. Stolypin's agrarian reform in Tsarist Russia in the 1900s, which aimed to change agricultural practices and reform the rural taxation system by implementing a land registry, collapsed in the face of joint resistance from the nobility and the peasants. In both cases, the State's efforts to make its environment more readable by simplifying it came up against the persistence of the very great complexity of ecosystems and practices.

Even those who renewed Harlesian cartography in the 2000s continued to focus almost exclusively on what "works", as Pickles (2004, p. 148) did, asking: "If maps precede and produce territories and social identities, what then are the objects and identities being produced in the digital transition? And what forms of territorialization are at work in new products of digital mapping?" At no point is the hypothesis that maps might have no effect, or effects very different from those predicted by theory (e.g. producing territories and identities) considered. Yet, weak signals about the existence of maps that do not reinforce domination effects sometimes appear almost accidentally. Thus, recent theses aimed at studying the effects and uses of geographic information technologies in various democratic arenas have led, almost unintentionally, to highlighting the non-effects and non-uses of these technologies. For example, Ritschard (2017), in looking at the role of spatial representations in integrated coastal zone management (ICZM) in Brittany, has multiplied the number of case studies questioning their non-use in the decision-making processes. A few years earlier, Mericskay (2013) made the same observation when he sought to analyze the uses of the geoweb and the Internet in the public debate through the observation of urban planning projects in Quebec City. This salutary research, with very unexpected results in relation to its initial objectives, is unfortunately still in the minority. There is still a lot of work to be done on all the cartographic projects that have been forgotten. However, one of our hypotheses is that with the digital transition, these cartographic failures, or these *maps for nothing*, are multiplying… but this is below the radar of the critics, who should, in our opinion, be looking for a possible political meaning to this phenomenon.

3.3.3. *A taste for great narratives*

On a level close to but different from the previous critique, we can also point to the persistent focus of critical cartography on globalizing histories of great cartographic projects. In many respects, this immoderate taste for the epic still irrigates many works within the movement. In his 2004 book, Jean Lefort describes the history of the concepts and mathematical tools of cartography as "a human adventure, that of scholars and explorers, from their dreams to the adventures of their voyages, from their vision of the world to the development of their instruments" (Lefort 2004, p. iv, authors' translation). These *great cartographic narratives* have constructed the identity of the discipline and may have given rise to vocations. It is interesting to observe that this fascination of researchers, which mainly concerned the cartographic exploits of conquerors or States, continues in relation to a collaborative and global project such as OpenStreetMap, which seduces by using a discourse that values disinterestedness, transnational volunteering and the capacity of a collective that has decided to cover the globe without professional prerequisites to give a quality results. The values of pioneers, discoverers, collective effort and *épopée* seem to be the driving force behind contemporary cartographic research today, regardless of the authors.

Thus, even if the recent history of digital mapping and GIS is less focused on the story of great explorations, it remains fascinated by the technological advances that shape its conception. Web documentaries, such as the *Geospatial Revolution* series produced by the University of Pennsylvania[7], are evidence of this focus on the technological feats that accompany contemporary mapping. By presenting applications in fields as diverse as logistics, crisis management and citizen science, this series highlights the widespread dissemination of geolocation techniques and the use of maps. Where the authors of this documentary see the massification of these tools as an epic, we see, on the contrary, an evolution that critics generally ignore: the map is no longer the sole domain of expert cartographers, and the diversification of stakeholders can be seen as revealing the end of the great cartographic narratives.

3.4. Reformulating critical mapping in the digital age

Following on from the various points and issues raised in the previous sections, this chapter offers some brief perspectives for an aggiornamento of critical cartography in the digital age, together with some examples of analysis relevant to current issues. Our main suggestion is that, when engaging in a political analysis of cartography in the broadest sense, we should suspend our propensity to presuppose

7 http://geospatialrevolution.psu.edu/.

from the outset that the map is an operator of power. In doing so, it seems to us to be much more heuristic to start from very empirical and embodied analyses, but also to be attentive to observing the effects of "maps" well beyond the actors who produce them, and the spaces they relate to. It is on this condition that their political effects cannot be overestimated, and that, on the other hand, weak signals can be detected, political effects that are not very visible but are very real. This approach, which could be described as prudent or parsimonious, requires disciplinary alliances to integrate the fine mechanics of the manufacture, uses and performativity of digital geographic information.

3.4.1. *Three priority disciplinary alliances*

Among the borrowings that are in our opinion necessary, those to be made from sociological institutionalism seem to us to be of primary importance. The work of Pierre Lascoumes and his colleagues working on environmental public action comes from the notion of *instruments*, inspired by Foucauld's study of power based on its daily *techniques.* They have opened up vast perspectives for understanding informational devices, which are also *instruments*. The latter can be conceived "as normative devices, both technical and social, with a generic vocation (applicable to a set of situations) and bearing a concrete conception of the governing/governed relationship based on a specific conception of regulation" (Lascoumes and Simard 2011, p. 18, authors' translation). It is immediately clear that this notion applies particularly well to the set of sociotechnical relations that surround the production of representations of space, today probably even more than before the digital era. It provides a good account of the multiple processes that make maps (in the broadest sense) have a decisive social effect in framing the problems affecting territories, and therefore also the solutions to be brought to them, an effect that is particularly clear in the field of environmental conflicts (Gautreau 2021). The other advantage of this approach is that it encourages observation of the political effects of mapping beyond or at the margin of the actors and spaces on which it focuses: what we call "instrument effects" are the "*original, and sometimes unexpected, effects that clearly exceed or stand out from initial expectations*" (Lascoumes 2004, authors' translation). This notion, close to that of "productivity", used by the sociology of environmental conflicts (Melé 2011), thus encourages attention to the multiple political effects of the contemporary cartographic factory, well beyond the mere relationship between a "producer" and a "receiver".

The second disciplinary alliance is to do with the insiders (Wood 2015, p. 16) of current cartography, that is, the designers of these systems. Critical cartographers are in fact no longer in a position to understand the functioning of the objects they are working on today without collaborating with computer scientists and geomatic experts, who alone are in a position where they are able to detail the lifecycle of spatial data in detail. When we know how important the technical details of a device (the size of the pixel in an image, for example) are for understanding its political effects (such as the invisibilization of certain resources in an image), we understand why this alliance is essential. As Dodge and Perkins (2015) point out, with the increasing place of algorithms in the production and circulation of representations of space, it is undoubtedly much trickier to deconstruct current cartographic devices, where diverse spheres of actors and multiple technologies are intertwined, than the devices of yesteryear where cartographic power was held by a few rare experts.

Moreover, cartographic order is now produced in part opaquely, from algorithmic black boxes, and by the invisible hands of those who feed them. To unfold this new factory and address concerns about the opacity of these algorithms (O'Neil 2016), the ethnographic approach allows us to take into account the entire ecosystem of algorithmic production and not have to solely rely on the analysis of computer code. Work on the proletarians of the digital economy (Casilli 2019), those "click workers" who, behind the illusion of artificial intelligence, multiply micro-tasks to drive the algorithms, has shown that the political effects of these systems cannot be interpreted solely from the intentions of their designers, but must incorporate the effects of thousands of invisible gestures. On the user side, it has also been shown that users are not simply influenced by the way the algorithm works, but also generate novel uses that were not intended by the designers (Bucher 2017). This approach amply deserves to be applied by critical cartography in the digital regime, as we believe it provides keys to understanding the current mutations of roles in social relations mediated by maps, and sheds light on their performativity.

3.4.2. *Making political sense of what does not work: investigating "maps for nothing" in the digital age*

In the early 2000s, the Bolivian municipalities that had obtained administrative autonomy in 1994 all adopted municipal land-use plans. With no in-house skills to produce them, and in a neoliberal context where the State had abandoned its role as the official cartographer, they resorted to receiving "help" from nature conservation NGOs, international cooperation agencies and private consultancy firms. The maps produced were highly accurate, and provided to the authorities in both paper and digital format, usually with a CD-ROM containing the geographic information layers generated by the cartographers. Ten years later, the vast majority of these maps had disappeared: either the municipal teams had forgotten the existence of such documents, or a faded paper printout of a map that no one consulted anymore remained in a corner of the office (Gautreau 2021).

Updating critical cartography in the digital age means, among other things, making political sense of such cartographic evaporation, seeking to understand why these maps are apparently produced *for nothing*. In neoliberal times, the ease with which digital developments can produce maps quickly has fostered truly unregulated markets for environmental mapping in developing countries. In the absence of a State that can standardize the production of maps and verify their quality *a posteriori*, many private and/or foreign actors have taken over this market and imposed their own standards, which are generally unsuited to the needs of municipalities. The latter, not having the means to negotiate with actors more powerful than themselves, therefore received maps of very limited use, or which the municipal teams did not have the means to interpret internally. But this evaporation is also linked to the material conditions of information storage, in poorly equipped premises, where the non-renewal of GIS software licenses quickly made the received layers unreadable, and where political instability led to the complete renewal of teams and acquired skills at each election, particularly in map reading (Gautreau 2021).

This example clearly shows that the political dimension of these phenomena of *cartography for nothing* does not shed light directly on the relations of domination mediated by maps, between municipalities and NGOs for example. But it does shed light on processes that are just as political, in this case on the functioning of development aid in South America in the 1990s or the abandonment of the State's claims to construct national representations of its territory (Lerch 2014). Here, digital technology plays on several contradictory levels that contribute, in the particular context of the exacerbated neoliberalism of that time, to slowing down the construction of a cartographic memory that the Bolivian population could capitalize democratically. It facilitates the creation of unregulated local cartographic markets, invested in by a large number of uncoordinated actors who are not concerned with the long-term conservation of their *deliverables*, and thus the short-term disappearance of a significant volume of maps, for which international cooperation has invested large sums. Collective memory, shared representations of space, increasing competence of lower administrative levels and cartographic dependence, *maps for nothing* shed light on all this.

3.4.3. *The political meaning of cartographic modesty: what do the "small maps of the Web" tell us?*

To shift the focus of the large State-owned mapping companies, we have proposed the expression "small maps of the Web" (Noucher 2017), which refers to the recent emergence, on the Internet, of diffuse practices that are multiplying exponentially: there have been thousands of maps made by activists, journalists, political actors, hackers, etc., that complement, bypass or compete with the production of official repositories. Historically, this notion thus makes it possible to signal this division, which partially marks the end of the great cartographic narratives, and testifies to the passage toward an era of cartographic post-sovereignty.

At the technical level, talking about small maps of the Web refers to the "small forms of the Web" (Candel *et al.* 2012, pp. 135–166); these new forms of writing have emerged on the Internet over the last 10 years, such as mashups, a characteristic Internet writing technique that aggregates content from various sites to create a new site. Among all the possible types of mashup (textual, iconographic, musical), *cartographic mashups* are the most frequent, in particular thanks to the success and ease of reuse of Google Maps. The maps that exploit these new possibilities are thus part of a renewal of the forms of geographical writing and are now fully part of the toolbox of Internet editorial writing. It is therefore advisable to analyze them not as isolated compositions but as elements of an editorial grammar whose framework of enunciation goes beyond the map alone; an observation that obliges critical cartography to take into account all the media interrelations of these maps, as opposed to considering them in isolation.

The expression also allows us to emphasize the political dimension of these new objects, by giving value to forms that were previously considered minor because they were produced outside the dominant institutional or commercial spheres. It is then a question of opening up the critical cartography to bodies of data that are currently ignored, of taking their political potential seriously, for example as forms of expression of minority groups. Shifting the gaze from "top-down" productions to understanding "bottom-up" map production makes it possible, for example, to make visible games of resistance sets or counter-hegemonic uses of cartography. This is not new, and counter-mapping projects (Peluso 1995) have long sought to subvert the power of maps. But this trend seems to have accelerated with the Internet to the point of occupying an important place in the media space, as in the case of the counter-maps that circulated in French Guyana to oppose the Montagne d'Or mining project (Noucher and Polidori 2020). It remains, of course, to measure their actual impact.

Finally, the small maps of the Web also refer to new forms of cartography. The intrusion of geolocation into everyday objects has gradually given rise to the figure of the "citizen as sensor" (Goodchild 2007), a connected individual who collects and disseminates, consciously or not, information about his environment or activities. Citizen science projects or the emergence of "self-quantification" testify to the success of these new ways of producing geographic information, generating personal maps (records of butterfly observations, tracks of bicycle rides, etc.). Located outside the paradigm of the map mediating a dominant/dominated relationship, these maps are not without political meaning, translating in their own way an individualistic and localist turn that can accompany certain digital practices. Thus, whether they come from militant community initiatives (counter-mapping) or from individual practices (ego-mapping), the small maps of the Web force critical cartography to renew its questions and methods.

3.4.4. *Circulation and policy*

In an interconnected world, the circulation of geographic information has now fully integrated the very production of maps. Thus, while geomatics has long functioned *in silos* with the multiplication, in the 1990s, of sectoral GIS (one GIS per territory and per theme), the interoperability that accompanies the development of the Web is now accelerating the networking of information. The simultaneous installation of numerous data infrastructures and their interconnection on a global scale allow the rapid constitution of a growing mass of information which, although scattered on different servers, can easily be gathered by a user and processed simultaneously.

Information system experts speak of orchestration to refer to the operations of controlling these data flows. These operations are sometimes taken for granted, as they are so often made invisible by automatic processes. However, it is precisely these orchestration logics that critical cartography should be concerned with today. Despite their progressive erasure from the cartographic scene, these logics act behind the scenes, both as vectors of standardization and as filters of representations deemed suitable or not for circulation.

In the space of a decade (2007–2017), the European Union has established a comprehensive project for the standardization of national and regional cartographic catalogs (from the way metadata should be filled in to the data model, from the functioning of consultation Web services to the graphic rules to be applied) with the implementation of the INSPIRE Directive[8]. In many respects, the versatility of these data infrastructures seems low, although they can theoretically accommodate a wide variety of representations of the territory (Gautreau and Noucher 2016). Yet, nothing is neutral in the choices of standardization or the criteria for selecting data flows: they are guided by principles, objectives and rules – in short, intentionalities that need to be deciphered. The circulation of spatial data thus appears to be a new form of control that homogenizes and filters representations, gradually becoming a tool for disseminating univocal representations of the territory, where alternative representations to those of the State or of the Web giants have difficulty finding their way. Henri Desbois underlines this paradox: "it is surprising to note that in our time when geomatics and the geoweb are at the origin of an inflation of cartographic production probably comparable by its magnitude to what Europe had experienced during the diffusion of the first printed atlases (Woodward 2007), we often all look at the same maps" (Desbois 2018, p. 321, authors' translation). In this way, collective visions are constructed and formatted that deserve to be discussed. Specifically geographical methods must therefore be invented to contribute to the understanding of these processes of decentralization, which in no way erase the effects of domination, creating or reinforcing hegemonic positions, of which Google is undoubtedly the most emblematic example.

While interoperability facilitates the transmission of data flows and makes it possible to spatially connect data sets (forming a *continent*), which could not previously interact and which evolved independently of each other (forming an *archipelago*), its implementation also has the corollary of plunging many data sets – small maps of the Web – into Internet limbo, for reasons that must also be questioned. For example, the analysis of the orchestration of data flows between different State platforms has made it possible to highlight the political and cultural reasons for the disappearance of Indigenous toponyms during the transition from Guianese infrastructures to French national infrastructures; these reasons being linked to issues of national sovereignty and to an imaginary world forged on the basis of past border conflicts (Noucher 2020). By multiplying this type of empirical study, by practicing a geography of the remains of the Internet or by remobilizing the blank parts of the maps in the era of the geoweb, it seems to us that critical cartography could go beyond its original fascinations around the finished product, cartographic successes or even great narratives.

8 Directive 2007/2/EC, known as the "INSPIRE Directive" of the European Parliament and of the Council of March 14, 2007 aims to ensure interoperability between databases and to facilitate the dissemination, availability, use and re-use of geographical information in Europe.

3.5. Conclusion

As the reader will have understood, this chapter raises many more questions than it answers about the challenges posed by the digital era to the critical cartographic movement. Indeed, we felt that it was a priority to first of all identify and prioritize the changes that have taken place in the very objects of this movement, and to propose an initial assessment of the situation; an assessment that is far from complete, given that digital technology is bringing about fundamental changes, many of which are still invisible. In the same way, this chapter has not proposed any "solutions", but has limited itself to giving examples of the way in which critical cartography can be "reformulated" now that the objects, devices, modes of circulation and reception are changing. As we have seen, the perspectives of aggiornamento do not consist of a theoretical and conceptual revolution, far from it: it is much more a question of decentering the gaze, of paying attention to bodies of work that were previously invisible, of looking for allies in order to decipher a cartographic factory that has become more complex. Today, politics is to be found in the details of multi-sited cartographic devices, in the economy of spatial representations, in apparently minor but massively used forms: it is up to the researchers to know how to spot them, abandoning certain antiphons of critical cartography, but preserving the essential achievements of a scientific movement that remains relevant today.

3.6. References

Anderson, B. (1991). *Imagined Communities*. Verso, London.

Bucher, T. (2017). The algorithmic imaginary: Exploring the ordinary affects of Facebook algorithms. *Information, Communication & Society*, 20(1), 30–44.

Candel, E., Jeanne Perrier, V., Souchier, E. (2012). Petites formes, grands desseins. D'une grammaire des énoncés éditoriaux à la standardisation des écritures. In *L'économie des écritures sur le web, volume 1 : Traces d'usage dans un corpus de sites de tourisme*, Davallon, J. (ed.). Hermès-Lavoisier, Cachan.

Casilli, A. (2019). *En attendant les robots. Enquête sur le travail du clic*. Le Seuil, Paris.

Crampton, J. (1994). Cartography's defining moment: The Peters projection controversy, 1974–1990. *Cartographica: The International Journal for Geographic Information and Geovisualization*, 31(4), 16–32.

Crampton, J. (2003). *The Political Mapping of Cyberspace*. Edinburgh University Press, Edinburgh.

Crampton, J. and Krygier, J. (2006). An introduction to critical cartography. *ACME: An International Journal for Critical Geographies*, 4(1), 11–33.

Desbois, H. (2015). La carte et le territoire à l'ère numérique. *Socio. La nouvelle revue des sciences sociales* [Online]. Available at: http://socio.revues.org/1262 [Accessed 7 June 2015].

Desbois, H. (2018). Conclusion : dans les coulisses du spectacle cartographique. *Comité français de cartographie*, 3(235–236), 321–324.

Dodge, M. and Perkins, C. (2015). Reflecting on J.B. Harley's influence and what he missed in "deconstructing the map". *Cartographica*, 50(1), 37–40.

Edney, M.H. (2019). *Cartography: The Ideal and Its History*. The University of Chicago Press, London.

Elwood, S. (2009). Geographic information science: New geovisualization technologies – Emerging questions and linkages with GIScience research. *Progress in Human Geography*, 33(2), 256–263.

Gautreau, P. (2021). *La Pachamama en bases de données. Géographie politique de l'information environnementale contemporaine*. Editions de l'IHEAL, Aubervilliers.

Gautreau, P. and Noucher, M. (2016). Sharing platforms in digital geographic information and spatial justice: Everything it promises? *Justice spatiale/Spatial Justice JSSJ* [Online]. Available at: http://www.jssj.org/article/information-geographique-numerique-et-justice-spatiale-les-promesses-du-partage/ [Accessed 29 November 2021].

Goodchild, M. (2007). Citizen as sensor: The world of volunteered geography. *GeoJournal*, 69(4), 211–221.

Goodchild, M. (2014). Two decades on: Critical GIScience since 1993. *The Canadian Geographer/Le Géographe canadien*, 59(1), 1–9.

Gould, P. and Bailly, A. (eds) (1995). *Le pouvoir des cartes. Brian Harley et la cartographie*. Anthropos, Paris.

Gruzinski, S. (1987). Colonial Indian maps in sixteenth-century Mexico: An essay in mixed cartography. *Res: Anthropology and Aesthetics*, 13(1), 46–61.

Gruzinski, S. (1993). *The Conquest of Mexico: The Incorporation of Indian Societies into the Western World, 16th–18th Centuries*. Polity Press, Oxford [Original published in French 1988].

Harley, J.B. (1989). Deconstructing the map. *Cartographica: The International Journal for Geographic Information and Geovisualization*, 26(2), 1–23.

Harvey, F.J. and Chrisman, N.R. (2004). The imbrication of geography and technology: The social construction of geographic information systems. In *Geography and Technology*, Brunn, S.D., Cutter, S.L., Harrington Jr., J.W., (eds). Kluwer Academic Publishers, Dordrecht.

Jasanoff, S. (2004). The Idiom of co-production. In *States of Knowledge: The Co-production of Science and the Social Order*, Jasanoff, S. (ed.). Routledge, London.

Joliveau, T., Noucher, M., Roche, S. (2013). Cartographie 2.0, vers une approche critique d'un nouveau régime cartographique. *L'Information géographique*, 77(4), 29–46.

Kain, R. and Baigent, E. (1992). *The Cadastral Map in the Service of the State. A History of Property Mapping*. The University of Chicago Press, Chicago.

Lascoumes, P. (2004). La Gouvernementalité : de la critique de l'État aux technologies du pouvoir. *Le portique*, 13(14), 169–190.

Lascoumes, P. and Simard, L. (2011). L'action publique au prisme de ses instruments. Introduction. *Revue française de science politique*, 61(1), 5–22.

Lefort, J. (2004). *L'aventure cartographique*. Belin, Paris.

Lerch, L. (2014). Mondialisation et digitalisation des territoires indigènes : rôles de l'aide internationale et des technologies de l'information géographique dans les politiques du territoire et du développement en Bolivie. PhD Thesis, Université de Genève, Geneva.

Melé, P. (2011). *Transactions territoriales. Patrimoine, environnement et actions collectives au Mexique*. Presses Universitaires François Rabelais, Tours.

Mericskay, B. (2013). Cartographie en ligne et planification participative. Analyse des usages du géoweb et d'Internet dans le débat public à travers le cas de la Ville de Québec. PhD Thesis, Université Laval, Quebec.

Monmonier, M. (1991). *How to Lie with Maps*. The University of Chicago Press. Chicago.

Mosco, V. (2014). *To the Cloud. Big Data in a Turbulent World*. Paradigm Publishers, Boulder.

Noucher, M. (2017). *Les petites cartes du web. Approches critique des nouvelles fabriques cartographiques*. Éditions rue d'Ulm, Presses de l'école normale supérieure, Paris.

Noucher, M. (2020). The place names of French Guiana in the face of the geoweb: Between data sovereignty, Indigenous knowledge, and cartographic deregulation. *Cartographica: The International Journal for Geographic Information and Geovisualization*, 55(1), 15–28.

Noucher, M. and Polidori, L. (eds) (2020). *Atlas critique de la Guyane*. CNRS Éditions, Paris.

O'Neil, C. (2016). *Weapons of Math Destruction. How Big Data Increases Inequality and Threatens Democracy*. Crown, New York.

Palsky, G. (2002). Philosophie de la carte [Review of *The New Nature of Maps. Essays in the History of Cartography*, by J.B. Harley, P. Laxton, and J.H. Andrews]. *L'Espace géographique*, 31(4), 375–377.

Palsky, G. (2003). L'esprit des cartes : approches historiques, sémiologiques et sociologiques en cartographie. HDR, Université Paris XII, Paris.

Peluso, N.L. (1995). Whose woods are these? Counter-mapping forest territories in Kalimantan, Indonesia. *Antipode*, 27(4), 383–406.

Pickles, J. (2004). *A History of Spaces: Cartographic Reason, Mapping and the Geo-coded World*. Taylor and Francis, London.

Rambaldi, G. (2005). Participatory 3-D modelling. In *Encyclopedia of Developing Regional Communities with Information Communication Technology*, Marshall, S., Taylor, W., Yu, X. (eds). IGI Global, Hershey.

Ritschard, L. (2017). Représentations spatiales et processus de Gestion Intégrée des Zones Côtières (GIZC) : application à deux territoires côtiers. PhD Thesis, Université de Bretagne Occidentale, Brest.

Rouvroy, A. and Stiegler, B. (2015). Le régime de vérité numérique. De la gouvernementalité algorithmique à un nouvel État de droit. *Socio. La nouvelle revue des sciences sociales*, (4), 113–140.

Rundstrom, R. (1991). Mapping postmodernism: Indigenous people and the changing direction of North American cartography. *Cartographica: The International Journal for Geographic Information and Geovisualization*, 28(2), 1–12.

Scott, J.C. (1998). *Seeing Like a State: How Certain Schemes to Improve the Human Condition Have Failed*. Yale University Press, New Haven.

Touzery, M. (2007). *De l'estime au cadastre en Europe : l'époque moderne*. Comité pour l'histoire économique et financière de la France, OpenEdition Books, Paris.

Trouillet, B. (2018). Les pêches dans la planification spatiale marine au crible des géotechnologies : perspectives critiques sur le "spatial" et "l'environnement". HDR, Université de Nantes, Nantes.

Wood, D. (1992). *The Power of Maps*. The Guilford Press, New York.

Wood, D. (2015). This is not about old maps. *Cartographica: The International Journal for Geographic Information and Geovisualization*, 50(1), 14–17.

Woodward, D. (ed.) (2007). *The History of Cartography, Volume 3: Cartography in the European Renaissance*. The University of Chicago Press, Chicago.

4

Mapping and Participation in the *Topos* and *Chora* Test

Federica BURINI
University of Bergamo, Italy

4.1. Introduction

This chapter analyzes the relationship between mapping and participation. It regards the former as a communication system to express the views of one or more actors in a specific territory, and addresses the latter within the complex framework of territorial governance, within which the political dimension of cartography is fundamental. To this end, this text adopts the reflexive and chorographic perspectives of semiotic approaches to cartography (Casti 2000, 2015): reflexivity is a way of interpretating and constructing cartographic representation, whereby the cartographer pays particular attention to their own role and actions; chorography considers the map as a complex system aimed at restoring the social and cultural meaning of the territory (the *chora*). The combination of reflexive and chorographic perspectives ensures the transparency of the cartographic production process, exploiting the potential of new digital technologies to represent the multiple points of view of the inhabitants, as well as cartographic techniques and languages that fall outside the ideals of topographic cartography (Casti 2015).

This text differs from studies that view participatory mapping as an instrument for collecting data from residents of a territory in order to better understand it. Instead, it focuses on the role played by multiple actors in the participatory mapping

For a color version of all the figures in this chapter, see www.iste.co.uk/debarbieux/mapping.zip.

The Politics of Mapping,
coordinated by Bernard DEBARBIEUX and Irène HIRT.

processes of producing and representing cartographic information. These actors are referred to as "inhabitants" in the sense of Lévy and Lussault. By focusing on inhabiting, they include all categories of individuals who use a given space and produce a spatiality, thus placing all actors on the same ontological level and on the same epistemological plane (Lévy and Lussault 2003, pp. 440–442). These inhabitants, who can become "cartographic interpreters", are experts as well as amateurs, residents and/or users of the represented territories. In this perspective, this contribution turns away from the map as a representation of material reality to focus on what it communicates about the relationship between inhabitants and territory. It considers the map as a social product that reflects the territorial knowledge-building practices of a given society, but also as an operator capable of conditioning actions in the territory. These dynamic processes of interaction and communication are fundamentally political.

We pay specific attention to the evolution of the forms and the media that make the relationship between cartography and participation possible. To do so, we will start from the analysis of historical maps, drawn on paper in contexts of colonial exploration and conquest, to then approach the collaborative cartographic systems that have emerged over the last few decades, focusing on the spatiotemporal analyses of mobility and displacement of inhabitants that took place thanks to these systems.

4.2. Participatory mapping tested by *topos* and *chora*

The term "participatory mapping" refers to communication systems that combine cartographic techniques and participatory methods, and whose objective is to take into account the spatial knowledge of the inhabitants of a given territory.

From its inception in the late 1980s, participatory mapping has been characterized by a plurality of actors who have given it its political tone: international development organizations, non-governmental organizations (NGOs), researchers, academics and members of local communities (local elected officials, administrative agents, representatives of associations, informal participants, etc.). The first forms of participatory mapping were practiced within rural development programs in the South. They were part of a "toolbox" of other participatory techniques capable of identifying the resources and sociospatial organization of a geographical area, in a short period of time, in order to carry out projects. In the early 1990s, the FAO (Food and Agriculture Organization of the United Nations) introduced the term "participatory mapping" in its Community Forestry Field Manual (FAO 1994, pp. 26–31). It justified the relevance of the participatory dimension because of its capacity to identify the local actors involved in a project and the social dynamics within a territory, and to mobilize their skills and traditional knowledge in natural resource management. Until then, such knowledge had been largely neglected, both by international organizations and by researchers involved in development projects.

The type of representation characteristic of these first forms of participatory mapping favors the topographic logic, inherited from the cartographic revolution of the early modern era. This logic corresponds mainly to the spatialization of information according to geometric reasoning, formalized by René Descartes in the 17th century; this logic emphasizes the measurement of territories and the representation of the material forms of which they are composed. This topographical logic owes its effectiveness to two main characteristics: (1) a cartographer, trained in the techniques of this type of representation, who carries an authority and a vision that is generally external to local communities; and (2) the collected data and a system of material objects selected by the same cartographer, who favors the analogical representation of sizes, positions and distances. However, by making cartography a functional means of expression, the topographic logic has led to a homogenizing vision of space that favors the point of view of institutional actors to the detriment of the symbolic and cultural representations of the territory of local communities. As an alternative, we propose to promote a chorographic logic of cartography, taking account of the relationships the territorial actors maintain with the inhabited, traversed and lived space (Casti 2015). This alternative extends the distinction between *topos* and *chora* theorized by Berque (2003): the former refers to the physical and absolute dimension of a place, while the latter refers to its narrative and relational dimension.

Without denying the importance of the referential function of the map, we propose here to demonstrate the added value of the chorographic logic in recent developments in participatory mapping. Such an approach makes it possible to take advantage, in the participatory process, of the presence of subjects who have a close link with the places represented, to mobilize their knowledge and experience, and to take into account the meanings and values they attach to these places. It is by taking note of the close relationship between society and territory, and of the richness of the social and cultural aspects of territorial reality, that participatory mapping reveals the political issues specific to the *chora*.

4.3. Toward a reflective approach to participatory mapping

In the literature, many terms refer to "participatory mapping", depending on the method of map production chosen and the geographic scale used. These terms include parish maps (Clifford and King 1996), maps resulting from "land games"

(Debarbieux and Lardon 2003; Lardon and Roche 2008), "community" maps and "Indigenous" maps (Hirt 2009; see also Chapter 7 of this book). More recently, they also refer to digital and online cartographies, such as geographic information systems (GIS), public participation GIS (PPGIS), community-based GIS (CGIS) (Craig *et al.* 2002), or collaborative and wiki systems (Roche *et al.* 2012).

The principle behind all of these forms of participatory mapping is to recognize the role of inhabitants as authors-mappers, which is no longer the exclusive privilege of professional cartographers. It also highlights the centrality of social actors in the creation of the map (individuals, communities, etc.), the diversity of methodologies, techniques and media used to encourage participation, as well as the variety of geographical scales of representation.

The semiotic approach has the advantage of highlighting iconization, namely the complex processes of communication through which the map, by its mere existence, and thanks to its self-referential power, presents conjectural facts as if they were truths (Casti 2015). Now, the reflexive and chorographic perspectives start from the assumption that map-producing actors are capable of questioning these "truths", adopting a critical stance and drawing methodological conclusions regarding the production and use of cartographic representation systems, including those mobilizing new digital technologies. These perspectives thus make it possible to better control the effects of the map in terms of communication, and to anticipate the social effects of its use to act on the territory.

They also lead us to prefer the term "collaborative mapping systems" to "participatory mapping": first, because "collaboration" between actors with diverse roles within a collective (Perkins 2007) suggests a more meaningful engagement on their part than "participation" (understood here as the inclusion of one actor in a mapping process planned by another actor and project) or "contribution" (understood here as the specific input or information given by an individual in a mapping project); second, because "collaborative systems" include a wider range of cartographic production techniques than "participatory systems", by including collaborative web-based cartographies, and the making of maps using algorithms powered by data produced by inhabitants (Burini 2016).

4.4. From solicited mapping to collaborative systems and digital traces

Cartographic practices in which the figure of the cartographic interpreter is embodied by the inhabitants are ancient. Concerning the period before the European conquests and colonization, these practices can be qualified as "Indigenous" in the etymological sense of the term, referring to the territorialization of social groups that originate from the very space in which they live. As part of a "self-centered" logic and an internal process of the group concerned, they contribute to social regulation by fixing certain symbolic and jurisdictional aspects of the territory on the map. This type of cartography, which cannot be defined as "participatory" because it is the result of a self-centered project, continues to be practiced today, as Book III of the series *The History of Cartography* points out, analyzing it in traditional African, American, Arctic, Australian and Pacific societies (Woodward and Lewis 1998).

It was European colonization – in other words, a fundamentally political fact – that led to the birth of the first forms of participatory mapping, which can be described as such because they were completed at the request of external actors for a specific project. Developed in the African context within the framework of "heterocentric" territorialization, these cartographies are described by Thomas Bassett as "solicited" (Bassett 1998, pp. 24–48). They originated from the need for European powers to benefit from local expertise and knowledge in order to know the conquered territories and their political organization, to orient themselves and move around, to exploit natural resources and to conduct commercial activities (Burini 2013). Figure 4.1 is an example of a solicited map made in 1885, 10 years before the war started in 1895 between the Kingdom of Italy and the Empire of Abyssinia (the former name of present-day Ethiopia), by an Abyssinian inhabitant at the request of the Italian explorer Antonio Cecchi. It was intended to enable the latter to understand how the villages in this part of East Africa were organized, in order to be able to imagine a project of conquest and colonization there.

From the end of the 1980s, a second type of cartography emerged that involved the participation of inhabitants at the request of an external actor: maps produced "for development", in the framework of projects implemented in the South (Burini 2012). The third form of participatory mapping refers to "collaborative digital systems", developed thanks to information and communication technologies (ICT) and the Internet. The dissemination of Participatory Geographic Information Systems (PPGIS or CGIS) has enabled local people to work with geomatics and computer experts and even to become experts themselves, able to use the systems easily and directly (Craig *et al.* 2002).

In short, for several decades now, the image of the leading cartographic interpreter has no longer been the preserve of the professional cartographer; it has gradually begun to include the inhabitant, whether a resident or user of the territory. Because of the interactivity of digital systems and high-speed Internet, the latter has become a key producer of cartographic information, useful for understanding the issues and meanings of a territory.

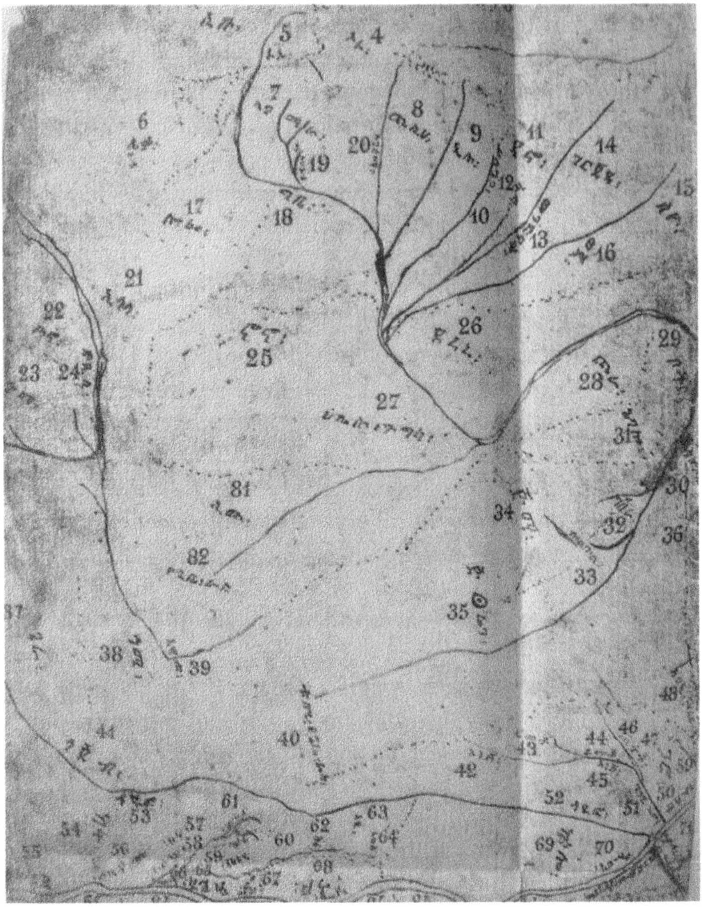

Figure 4.1. *Map drawn by an Abyssinian, requested by an Italian explorer (source: Library of the Italian Geographical Society in Rome, Cecchi 1886 (p. 551))*

In the last decade, in the context of "data-driven science" (Kitchin 2014) and the increasing flow of data of a geographical nature (big geodata) (Graham and Shelton 2013), a novel form of mapping involving the inhabitants of a territory has developed. This involves the representation of "digital traces", produced actively or passively by "citizens as sensors", so named by Goodchild (2007) by analogy with fixed or mobile sensors (p. 212). These individuals produce georeferenced data simply by owning a device connected to a global positioning system (GPS) sensor or cell phone, or by sharing an experience on the Web or via social networks. These devices measure localized phenomena, such as the flow of people. The term "sensor", however, reflects Goodchild's technocratic conception of the reconfiguration of the individual's role in these new forms of mapping; a conception that largely underestimates their social and territorial implications.

At the turn of the new millennium, thanks to Geoweb 2.0, this evolution in the production of georeferenced data was accompanied by significant technological innovations that facilitated the interactivity of online digital mapping. This cartographic revolution was born out of the transition from paper to digital until its introduction on the Web. The central role of the inhabitant in these interactive modalities is unprecedented both in terms of the collection of data and in terms of the cartographic production process. According to Lévy (2015), this "cartographic turn" represents a real paradigm shift that affects other domains, that is, "a turn of turns": spatial, pragmatic, linguistic, ethical and digital (pp. 15–17).

4.5. The new boundaries of collaborative mapping systems produced by geolocation technologies

Approaches based exclusively on topographic metrics – that is, on a Euclidean geometric method of measuring distance, or on static data from inhabitants, such as demographic data from censuses – do not capture the characteristics of space in the current context of globalization (Lévy 2008). An analysis of the contemporary world requires topological metrics, capable of representing a discontinuous space with fluid boundaries, constantly redefined by individuals who assign increasingly flexible statuses and functions to places (public, semi-public, private, semi-private), and create nodes and connections in a network without a beginning or an end in time and space. In order to measure and understand these topological metrics, it is necessary to investigate the experiences, views and skills of users, and to identify the new data sets they generate. It is necessary to know the location of users, their movements and methods of travel, and to reconstruct their use and experience of places.

As for the possibility of collecting data relating to the geolocation of individuals – in other words, information of a referential and denotative nature – it is made possible by the rapid technological progress (GPS, smartphones, sensor networks, cloud computing, etc.), which have radically transformed the way in which geographical data are collected, stored, disseminated, analyzed, visualized and used. Indeed, these technologies make it possible to restore the spatiotemporal dimension of a phenomenon in two ways: on the one hand, through collaborative systems allowing individuals to consciously and voluntarily express (via social networks, blogs, websites, applications, collaborative mapping platforms) their experience of a space (judgments, practices, needs, perspectives) by communicating information that includes not only geographic coordinates but also toponyms, semantic categories or any other element that allows for characterizing a place; on the other hand, through sensors associated with objects or living beings (human and animals), in motion or placed at a fixed point. As mentioned previously, sensors associated with an individual often imply a passive role on their part. The user does not necessarily consciously reveal their geolocation, provided by the tools they use (computer, tablet, smartphone, GPS, etc.). But whether their participation is voluntary or not, the data produced have considerable potential for the study of geographical space: (1) they allow reconstruction of the movements of the inhabitants; and (2) they are produced by the individual and show their relationship with the places.

The main production areas for these new data are urban. Residents, commuters, tourists and migrants are all users of these urban nodes which, according to the United Nations, are home to 60% of the world's population. Moreover, these urban areas concentrate the telecommunication infrastructures necessary for data production. Metropolises, as well as small and medium-sized cities, are investing in infrastructures that promote ever-increasing spread of high-speed Internet (also via wireless network hot-spots), and in a system of sensors collecting data related to the different sectors of urbanism (urban traffic, pollution, waste management, tourism, etc.) (Batty 2013). As evidenced by the article "Big Data and the city" in the journal *Built Environment,* cities are now laboratories for the collection, monitoring and analysis of big geodata: the revolution brought about by big geodata in the field of urban governance lies in the spatiotemporal dimension of the data produced by the users of space (Batty 2016).

4.6. Representing the spatiotemporal dimension of urban mobility using data produced by inhabitants

Until a few years ago, it was difficult for a geographer to have access to a large amount of data generated by inhabitants over a given period of time; data relating to their experiences and perceptions of places, their itineraries and their movements in the urban space. The analysis of the flow of people had to be limited to data provided by statistical agencies, possibly combined with an extended field survey. It could also be based on data from telephone companies, provided that researchers could afford the cost of acquiring them. Today, these data are accessible because of the level of connectedness of urban places from connections activated by individuals in the places they visit and information shared on the Web. Lévy uses the notion of the rhizome to define this "relational space of an individual" or this "family of networks characterized by the absence of identifiable boundaries" (Lévy 2014, p. 49). It is therefore the space produced by individual action and experience via mobility, and by the multifaceted relationships established between individuals, mobile or not, woven through telecommunications (Maitre *et al.* 2016, p. 21).

There are many studies on these urban spatialities in the literature. For example, those related to spatiotemporal journeys made by tourists, reconstructed due to social media and data associated with images uploaded by individuals (Girardin *et al.* 2008; Agryzkov *et al.* 2016). This is the case of the cartographic experiments of the "data-artist" Eric Fischer uses Twitter and Flickr to identify the routes of tourists or residents of Paris, London, Manchester, Moscow and other European and non-European cities[1]. However, these studies are topographic in nature, both in the analysis of the quantitative data collected and in the type of cartographic representation proposed. The analyses are based on simple geolocations, generally from direct data, or by geolocating Tweets. Indirect data (e.g., related to the toponyms of the places mentioned in the Tweet) are rarely generated, and it is also impossible to display the descriptions, multimedia metadata (photos, videos, etc.) of the visited places for each inhabitant and their connections.

Therefore, these data, although useful for understanding the spatial distribution of phenomena on an urban scale, provide little relevant information for a reflexive and chorographic approach. They do not allow us to reconstruct the behavior of individuals and the meaning that individuals give to places, following the example of García-Palomares *et al.* (2015, p. 410), who identify the spaces frequented or traversed, the time spent in them and the means of transport used. For this purpose, voluntary geographic information (VGI) (Goodchild 2007; Elwood *et al.* 2012; Sui *et al.* 2013), produced notably through the contribution of individuals to social media, should also provide information on the experience of places of the inhabitants. These enable a new interpretation of urban mobility by shifting the analysis from the macro-level to the level of "small data", that is, those capable of restoring the point of view of individuals, or showing the space of individual actions drawn by their mobility and rhizome. This does not necessarily mean shifting to a larger scale, that is, a smaller territory, but rather considering the passage made by the individual between the different scales according to the means of transport used.

A critical approach to understanding urban mobility using big geodata, however, requires that geographers go beyond simply collecting and mapping the information produced by the inhabitants. It must also include a reflection on the spatial capital of individuals, a term introduced by Lévy to complement the three dimensions – economic, cultural and relational – of Bourdieu's concept of social capital. According to Lévy, spatial capital consists of "a stock of geographical experiences

1 See: https://labs.mapbox.com/labs/twitter-gnip/locals/#3/10.49/49.75.

transformed into heritage and a flow of skills capable of making this heritage bear fruit by mastering places and the links with these places" (Lévy 2003, p. 215). Identifying the spatial capital of individuals through big geodata thus amounts to showing the specificity of the places frequented, the nodes of mobility, the role of cities as primary or secondary destinations on the basis of the information, comments and opinions generated by those who visit them (Burini 2014).

Big geodata can play an important role in the analysis of urbanity *a posteriori*. Indeed, the observation of the city from the data produced by its users allows for knowing and understanding certain phenomena but also managing them and facilitating decision-making. Such an analysis is becoming increasingly relevant in the field of mobility and urban planning, especially to measure the gap between existing infrastructures and the needs of inhabitants (Beaude 2015, pp. 283–285). In this sense, it constitutes a common good as it promotes better urban governance, attentive to the real needs of those who frequent cities, even if they are not the residents.

4.7. Toward a reflective and chorographic approach to collaborative mapping systems

In this section, we demonstrate the potential of the reflexive and chorographic perspectives, both methodologically and empirically, and their applicability in different urban and non-urban contexts. Based on the main elements of chorographic semiosis, we propose a typology of collaborative cartographic systems according to the role of the *cartographic interpreter* (individual or collective actors) in the production of cartographic information, and the active or passive form of their participation (Table 4.1).

Collaborative mapping systems with active participation include systems that are both *solicited* and *voluntary*. They are implemented by inhabitants actively engaged in the collection of data considered necessary to act on the territory or to produce territorial knowledge. The so-called *derived* systems involve the *passive* participation of inhabitants in the production of georeferenced digital traces (Girardin *et al.* 2008; Beaude 2015); consciously or not, these transmit information by the mere fact of being connected to the Web or to a telecommunication service with a geolocation device.

Examining the three cartographic systems – solicited, voluntary and derivative – from a semiotic point of view allows us to analyze them as complex communication systems.

Inhabitant participation/ mapping process	Active		Passive
	Solicited	Voluntary	Derived
Cartographic interpreter	Collaboration between actors involved in a territorial project and inhabitants solicited	Web users participating in a collective mapping project on a voluntary basis	Experts produce a map from georeferenced digital traces transmitted by Web users or "citizen sensors"
Data collection method	Collection carried out individually or collectively, at the request of an external actor	Voluntary and individual contributions to a mapping system for collective use	Derived collection of individual digital and georeferenced traces of the inhabitants (cell phones, mobile apps or services, social networks, Wi-Fi, Bluetooth terminals)
Information content	Information on the territorial resources present in a territory: accessibility, use and organization; skills and knowledge of the inhabitants, conflicting interests, etc.	Places and routes experienced by an individual	Traces left by the individual's connectivity (travel, catering, tourism, community events, sports activities, etc.)
Objectives of the mapping process	Territorial development project, planning project	Production of knowledge of collective interest; solving problems of spatial accessibility or rescue in times of crisis	Representing the spatiotemporal dimension of the digital traces produced by the inhabitants
Metrics and graphic transposition	Topography/ chorography	Topography	Topography/ chorography
Pragmatic potential	Decision-making on land-use planning based on the needs of inhabitants	Accessibility to the territory, organization of emergency services, resolution of environmental problems, etc.	Spatiality *a posteriori*: the movements of individuals generate new knowledge about the use of places
Examples	Participatory maps, cybercartography	OpenStreetMap (OSM), Wiki-GIS, crisis mapping, citizen science	Mapping of social networks, the Web, heat-maps generated by big geodata

Table 4.1. *Collaborative mapping systems and levels of inhabitant participation*

4.7.1. *Solicited mapping systems with active participation*

In the case of mapping systems that are solicited with active participation, there are multiple cartographic interpreters collaborating in the production of cartographic information: on the one hand, the various actors participating in a territorial project and, on the other hand, the different categories of inhabitants solicited (resource persons or referents of a community, representatives of certain professional categories or groups of inhabitants by age, gender, typology of users of the city, etc.). The participation of the inhabitants, whether individual or collective, in the production of the map (on paper or in digital format) is direct and takes place at the request of external actors. These mapping systems concern areas identified by the stakeholders of a territorial planning or development project, within which the inhabitants concerned transmit knowledge on aspects relating to territorial resources: accessibility, use and organization of resources, skills and knowledge of the stakeholders on the latter, conflicting interests, etc. Although the motivations leading to the implementation of such mapping systems are varied, the objective is invariably to promote local governance and citizen or community participation.

Figure 4.2 shows a map solicited in the framework of an environmental cooperation project in the peripheral areas of the W-RBTW Transboundary Biosphere Reserve (Benin, Burkina Faso, Niger), and of participatory research conducted by the Diathesis Cartographic Laboratory of the University of Bergamo between 2002 and 2006 (Burini 2016). The map was first drawn on paper by members of the fishermen's association of the village of Bossia in Niger at the request of the researchers, in order to highlight the wealth of their knowledge on the use of water resources in the villages and peripheries of the W-RBTW. The data collected were then transposed and georeferenced in a community GIS on an aerial background map, with the aim of producing a readable and easily understandable topographic representation for the managers of the protected area. The political significance of this document is high, as it enabled the fishermen's association to negotiate with managers for seasonal use of the rivers, and thus gain access to natural resources that they had been deprived of since the creation of the Park in colonial times.

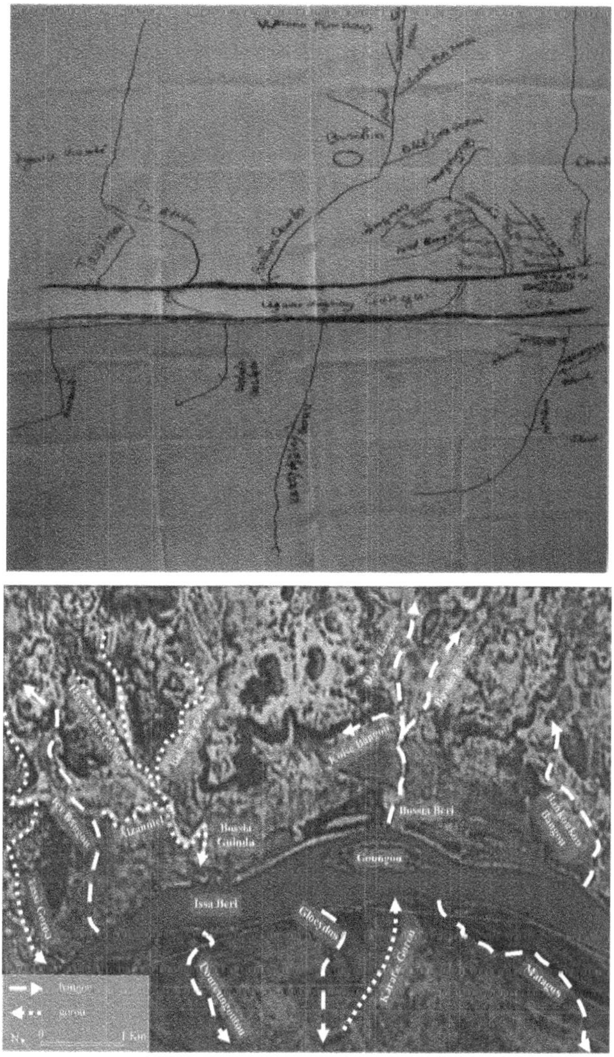

Figure 4.2. *The requested mapping system of the Bossia (Niger) hydrographic system (source: map drawn by F. Burini and Bossia villagers, in Burini 2006)*

Figure 4.3 is an example of an active participation mapping system available on the Web. This map is the result of a territorial governance project carried out in the context of the candidacy of the city of Bergamo for the status of "European Capital of Culture". Supported by the municipality, the cartographic project was implemented by the Diathesis Laboratory of the University of Bergamo through a participatory process that included – in addition to consulting the inhabitants on the Web – meetings and focus groups to collect cultural initiatives to be promoted during the year of the bid. The project allowed for the representation of proposals from associations or citizens concerning the cultural enhancement of natural resources or little-known places in the territory of Bergamo.

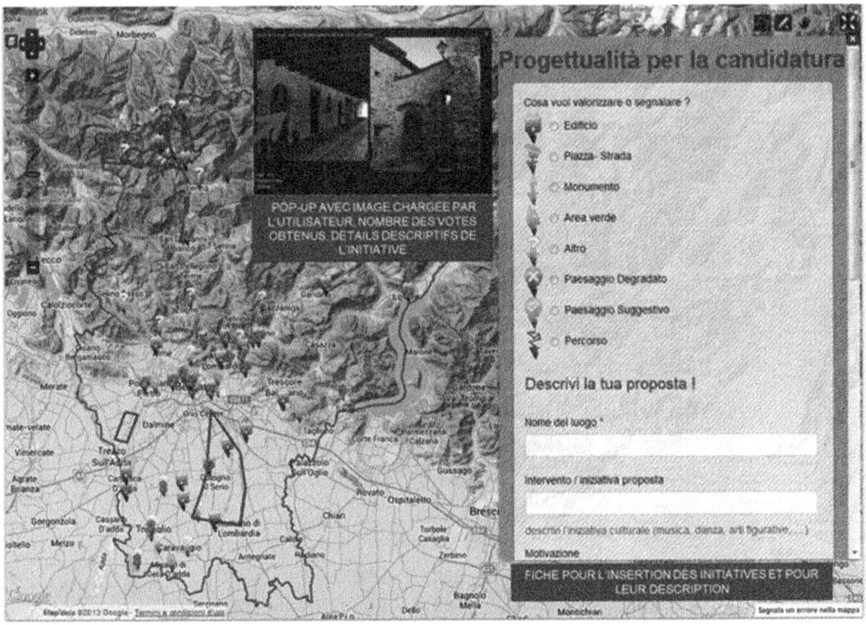

Figure 4.3. *The Bergamo Open Mapping system requested for Bergamo's application for the "European Capital of Culture" title (source: map by DiathesisLab for the University of Bergamo, 2013)*

These two examples show that the graphical transposition of solicited cartographic systems varies according to the "evolution of ICT, and that it can be part of a topographic and/or chorographic metric logic. They also suggest that solicited mapping systems with active participation have political significance. Their pragmatic potential for the optimization of territorial governance is therefore important, although it is not always taken into account by the various actors who solicit participatory paths. In particular, these systems have the power to orient the decision-making of project promoters toward the needs of the inhabitants expressed on the map and consequently to allow recognition of the spatial capital of the communities that participate in developing the maps.

The most advanced example of this type of mapping today is represented by "cybercartography" (Taylor 2005; Taylor and Lauriault 2014; Taylor *et al.* 2019), developed by the Geomatics and Cartographic Research Centre at Carleton University in Ottawa, directed by Fraser Taylor[2]. The specific purpose of cybercartography is the creation of collaborative multi-sensory and multimedia Web-based mapping systems that can integrate the narratives, journeys and knowledge of Canada's Indigenous communities through multimedia systems (see Chapter 7 of this book).

4.7.2. *Voluntary and participatory mapping systems*

In voluntary and participatory mapping systems, residents voluntarily produce georeferenced data. As Web users, they share their knowledge or skills within a project of a collective nature. Because of this voluntary dimension, it is scarcely possible to know *a priori* the number and categories of inhabitants who decide to participate in a project. By providing information about places and routes they know, Web users transform their personal experience into a collective cartographic heritage for other inhabitants, especially those who have access to the relevant mapping system. If this information is used by governmental or institutional actors, its collective value is increased.

Voluntary and participatory mapping systems meet several objectives: first, their aim may be to produce knowledge of collective interest to compensate for the absence or inadequacy of cartographic information on a given area. This is the case of OpenStreetMap (OSM), invented in 2004, which presents itself as the world's[3] first free geographic database. Using the GPS system and open access data, OSM is part of geomatics 2.0, VGI and "neogeography"[4] (Turner 2006; Wilson and Graham 2013). Second, some of these systems are intended to respond to rapidly changing conditions of organization and displacement in territories in crisis (e.g. during a natural disaster or humanitarian crisis), hence the name "crisis mapping" given to this type of device (Burini 2014). Third, these systems often support citizen science processes, providing useful data for monitoring environmental problems, for example.

---

2 See: https://gcrc.carleton.ca/index.html?module=module.gcrcatlas_cybercartography.

3 The number of contributors currently participating in OpenStreetMap is almost 7 million, see: www.openstreetmap.org.

4 The concept, coined by developer Andrew J. Turner, refers to "a set of techniques and tools that fall outside the domain of traditional GIS" (Turner 2006). Other authors have also used this concept (Rana and Joliveau 2009; Liu and Palen 2010; Warf and Sui 2010).

The interactivity of voluntary participatory mapping systems is generally limited. Furthermore, the user is not able to change the map symbols, chosen by the cartographer, for reasons of standardization of geographic information. Finally, the location of places is considered more important information than their social meaning, which is not included in the graphic representation. The Spanish platform *Frena la Curva* ("slow down the curve") is a good example of the type of interactivity that is possible (see Figure 4.4). This crowdsourcing device mobilizes the open-source Ushahidi[5] system, developed in Kenya to map the post-election violence of 2008. It has since been used by thousands of people around the world, especially during natural disasters and political crises. *Frena la Curva* is a citizen and neighborhood support platform that was created in Spain on March 19, 2020, as an emergency response to the Covid-19 pandemic.

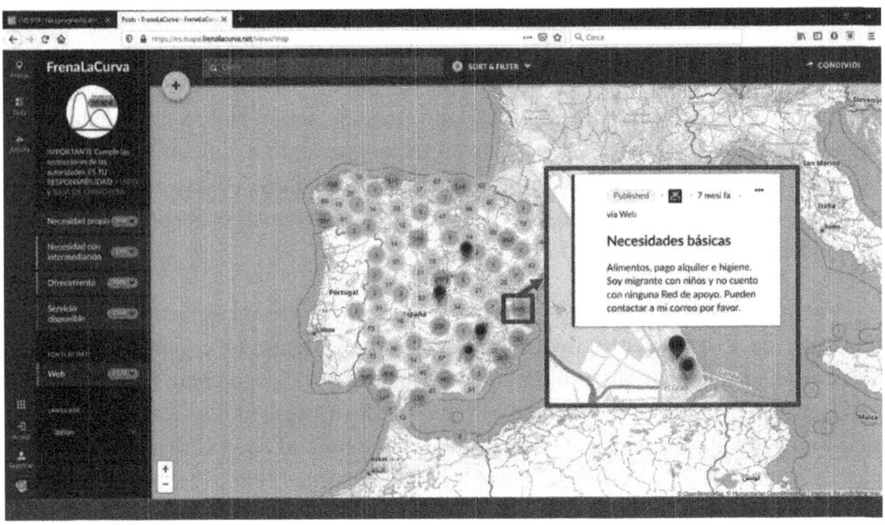

Figure 4.4. *The voluntary Ushahidi Frena La Curva system for Covid-19, released in March 2020 (an icon with the total number of alerts in a given area, and individual icons are highlighted) (source: https://es.mapa.frenalacurva.net/views/map)*

The user can identify and geolocate information by selecting four categories of icons: (1) red icons: request for care, food or other primary goods and services; (2) orange icons: the same type of request but formulated through a third party; (3) green icons: offer of services; (4) blue icons: offer of care, food or other primary goods and services by public or private bodies, or associations. The interactivity of

5 Ushahidi is the Swahili word for "testimony". See: www.ushahidi.com.

the system allows users to insert textual descriptions to detail their requests or the offer of a service. The geolocation of more than 9,000 places – observed in mid-November 2020 – underlines the importance of the role of collaborative mapping systems in crisis situations. If the user does not have the option of modifying the icons and the colors attributed to the geolocated points, it is because the main objective of *Frena la Curva* is to collect referential information in the simplest way possible, giving priority to the location criterion. It is therefore easy to know the number of reports created by users for each city or territory due to icons indicating the total number of contributions for each place. On the other hand, access to more detailed information – in other words, information that can express the spatial capital of the inhabitants – involves clicking on each icon indicating a total number of reports. This allows for moving to a larger scale relative to a smaller area and so on, until the individual content of a single user's icon on a given location can be accessed. Voluntary cartographic systems thus promote the topographic logic (location, quantification of alerts) to the detriment of the chorographic logic (content and information on the spatial capital of the inhabitants). They thus reinforce the knowledge of the location of certain places, routes or infrastructures by providing information related to the accessibility of the territory or the organization of emergency services. To date, however, these systems do not provide for a more detailed automatic elaboration of VGI in order to create icons that can reflect the spatial capital of individuals and their experience of places.

4.7.3. *Derivative and passive participatory map systems*

The data in derivative passive mapping systems are derived from the "traces" left by individuals on the Web as a result of their connectivity. They are related to sports, travel, dining and tourism activities. They are derived either from information posted by people wishing to share an experience of a place on social networks (texts, photos, videos), or involuntarily by "sensor" individuals, leaving digital traces of their activities on the Web, consciously or not. They thus become agents contributing – albeit indirectly – to the production of new representations of the spatiotemporal dimension of individual mobility.

Although these georeferenced data are produced by amateurs, their processing and analysis involves the intervention of experts – geographers, geomaticians and computer scientists. And even though some of this information from amateurs is free of explicitly spatial content (searching for information on Internet browsers, writing texts on blogs, etc.), experts can map it, thus producing new content without any necessary link to the initial objectives of this information. An example is the TweetMap-derived mapping system produced by the Diathesis Laboratory of the

University of Bergamo, as part of the Urban Nexus project. In this example, Twitter and Instagram data of the user inhabitants of the "Il Caravaggio" airport, near Bergamo (Italy) were used (Burini *et al.* 2018, pp. 111–130) (see Figure 4.5).

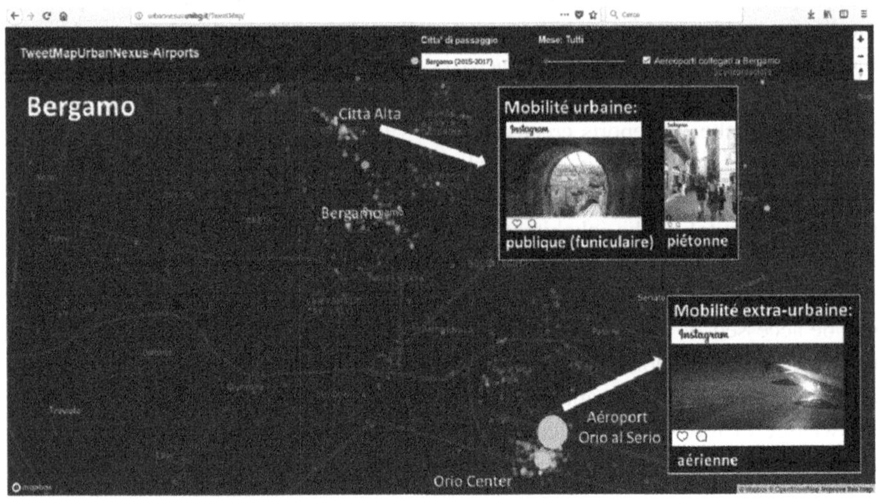

Figure 4.5. *The TweetMap-derived mapping system (source: CST-DiathesisLab for the Urban Nexus project, in Burini (2019))*

Derived mapping systems are potentially useful for territorial governance because they reveal *a posteriori* spatiality: they thus highlight how the places are used, the rhythm of the territories – that is, the progressive concentration of inhabitants in specific places and times – as well as the real and virtual connectivity between territories, when they are actually traveled through, or when they are searched for on the Web by several people at a time.

4.8. Conclusion

Semiotic approaches to collaborative cartographic systems show that the relationship between cartography and participation is characterized by an unprecedented wealth of cartographic modalities and information sources. This abundance is the result of dynamic processes of co-production of information by a plurality of territorial actors, and the unprecedented possibilities of mapping the spatialities of individuals *a posteriori*, because of the data produced by the inhabitants themselves and the capacity of researchers to find new sources of data in an interdisciplinary manner, to collect them and to locate them on a map. This growing availability of data constitutes an important potential form of support for decision-making authorities in territorial governance, in the fields of administration, urban or territorial planning and environmental management.

However, the cartographic challenge is significant: it is not only a question of bringing together, in the same cartographic system, the different forms of participation of the inhabitants in the production of data but also of developing the technical capacity to render in the same cartographic system the *topos* and the *chora*. As we have shown in this chapter, topography is the most used logic in the field of participatory mapping (solicited, voluntary or derived). However, behind the cartographic icons lies a world of spatial knowledge (beyond simple geolocation) that it is time to reveal as spatial capital of a chorographic nature.

4.9. References

Agryzkov, T., Marti, P., Nolasco-Cirugeda, A., Serrano-Estrada, L., Tortosa, L., Vicent, J.-F. (2016). Analysing successful public spaces in an urban street network using data from the social networks Foursquare and Twitter. *Applied Network Science*, 1(1), 12.

Bassett, T. (1998). Indigenous mapmaking in intertropical Africa. In *The History of Cartography: Cartography in the Traditional African, American, Arctic, Australian, and Pacific Societies*, Woodward, D. and Lewis, M. (eds). University of Chicago Press, Chicago.

Batty, M. (2013). Big data, smart cities, and city planning. *Dialogues in Human Geography*, 3, 274–279.

Batty, M. (2016). Big data and the city. *Built Environment*, 42(3), 321–337.

Beaude, B. (2015). Lost in transduction: From digital footprints to urbanity. In *A Cartographic Turn: Mapping and the Spatial Challenges in Social Sciences*, Lévy, J. (ed.). EPFL Press/Routledge, Lausanne.

Berque, A. (2003). "Lieu" 1 [Online]. Available at: https://www.espacestemps.net/articles/lieu-def1/ [Accessed 1 December 2021].

Burini, F. (2006). La cartografia partecipativa e la cooperazione ambientale in Africa: Il caso di Bossia (Niger). *Bollettino della Società Geografica Italiana*, 11(4), 977–996.

Burini, F. (2012). Cartographie et participation pour la coopération environnementale : le terrain et la restitution des savoirs traditionnels en Afrique subsaharienne. *Annales de géographie*, 687–688(5–6), 487–512.

Burini, F. (2013). L'évolution de la cartographie auprès des sociétés traditionnelles en Afrique subsaharienne. *L'information géographique*, 77(4), 68–87.

Burini, F. (2014). Le capital spatial dans "un tweet" ? Le crisis mapping en contextes urbains [Online]. Available at: http://www.espacestemps.net/en/articles/le-capital-spatial-dans-un-tweet-le-crisis-mapping-en-contextes-urbains/.

Burini, F. (2016). *Cartografia partecipativa. Mapping per la governance ambientale e urbana*. Franco Angeli, Milan.

Burini, F. (2019). Sistemi partecipativi per la governance territoriale: Cartografare i dati prodotti dagli abitanti. In *La Geografia a Bergamo. Nuove sfide per l'analisi territoriale e il mapping*, Casti, E. (ed.). AGEI, Rome.

Burini, F., Ciriello, D.E., Ghisalberti, A., Psaila, G. (2018). The Urban Nexus project: When urban mobility analysis, VGI and data science meet together. In *Mobile Information Systems Leveraging Volunteered Geographic Information for Earth Observation: Earth Systems Data and Models*, Bordogna, G. and Carrara, P. (eds). Springer International Publishing, Cham.

Casti, E. (2000). *Reality as Representation*. Bergamo University Press, Bergamo.

Casti, E. (2015). *Reflexive Cartography: A Modern Perspective in Mapping*. Elsevier, Amsterdam.

Cecchi, A. (1886). *Da Zeila alle frontiere del Caffa*. Loesher, Rome.

Clifford, S. and King, A. (1996). *From Place to PLACE: Maps and Parish Maps*. Common Ground, London.

Craig, W., Harris, T., Weiner, D. (2002). *Community Participation and Geographic Information Systems*. Taylor & Francis, London.

Debarbieux, B. and Lardon, S. (2003). *Les figures du projet territorial*. Editions de l'Aube/DATAR, Paris.

Elwood, S., Goodchild, M.F., Sui, D.Z. (2012). Researching volunteered geographic information: Spatial data, geographic research, and new social practice. *Ann. Assoc. Am. Geogr.*, 102, 571–590.

FAO (1994). Community forestry field manual. Tree and land tenure: Rapid appraisal tools. Manual, Food and Agriculture Organization, Rome.

García-Palomares, J.C., Gutierrez, J., Minguez, C. (2015). Identification of tourist hot spots based on social networks: A comparative analysis of European metropolises using photo-sharing services and GIS. *Applied Geography*, 63, 408–417.

Girardin, F., Calabrese, F., Fiore, F.D., Ratti, C., Blat, J. (2008). Digital footprinting: Uncovering tourists with user-generated content. *IEEE Pervasive Computing*, 7(4), 36–43.

Goodchild, M.F. (2007). Citizens as sensors: The world of volunteered geography. *GeoJournal*, 69(4), 211–221.

Graham, M. and Shelton, T. (2013). Geography and the future of big data, big data and the future of geography. *Dialogues in Human Geography*, 3(3), 255–261.

Hirt, I. (2009). Cartographies autochtones. Éléments pour une analyse critique. *L'Espace géographique*, 38(2), 171–186.

Kitchin, R. (2014). Big Data, new epistemologies and paradigm shifts. *Big Data & Society*, 1–12.

Lardon, S. and Roche, S. (2008). Représentations spatiales dans les démarches participatives. *Revue Internationale de Géomatique*, 18(4), 423–572.

Lévy, J. (2003). Capital spatial. In *Dictionnaire de la géographie et de l'espace des sociétés*, Lévy, J. and Lussault, M. (eds). Belin, Paris.

Lévy, J. (2008). *L'invention du monde. Une géographie de la mondialisation.* Presses des Sciences Po, Paris.

Lévy, J. (2013). *Réinventer la France : trente cartes pour une nouvelle géographie*, Fayard, Paris.

Lévy, J. (2014). Inhabiting. In *The Sage Handbook of Human Geography*, Lee, R., Castree, N., Kitchin, R., Lawson, V., Paasi, A. (eds). Sage, London.

Lévy, J. (2015). *A Cartographic Turn: Mapping and the Spatial Challenges in Social Sciences.* EPFL Press/Routledge, Lausanne.

Lévy, J. and Lussault, M. (2003). Habiter. In *Dictionnaire de la géographie et de l'espace des sociétés*, Lévy, J. and Lussault, M. (eds). Belin, Paris.

Liu, S. and Palen, L. (2010). The new cartographers: Crisis map mashups and the emergence of neogeographic practices. *Cartographic and Geographic Information Science*, 37(1), 69–90.

Maitre, O., Lévy, J., Romany, T. (2016). Rebattre les cartes. Topographie et topologie dans la cartographie contemporaine. *Réseaux*, 1(195), 17–52.

Perkins, C. (2007). Community mapping. *The Cartographic Journal*, 44(2), 127–137.

Rana, S. and Joliveau, T. (2009). NeoGeography: An extension of mainstream geography for everyone made by everyone? *Journal of Location Based Services*, 3(2), 75–81.

Roche, S., Mericskay, B., Batita, W., Bach, M., Rondeau, M. (2012). WikiGIS basic concepts: Web 2.0 for geospatial collaboration. *Future Internet*, 4(1), 265–284.

Sui, D., Elwood, S., Goodchild, M. (2013). *Crowdsourcing Geographic Knowledge: Volunteered Geographic Information (VGI) in Theory and Practice.* Springer, Dordrecht.

Taylor, D.R.F. (2005). *Cybercartography: Theory and Practice.* Elsevier, Amsterdam.

Taylor, D.R.F. and Lauriault, T.P. (2014). *Developments in the Theory and Practice of Cybercartography.* Elsevier, Amsterdam.

Taylor, D.R.F., Anonby, E., Murasugi, K. (eds) (2019). *Further Developments in the Theory and Practice of Cybercartography. International Dimensions and Language Mapping.* Elsevier, Amsterdam.

Turner, A. (2006). *Introduction to Neogeography.* O'Reilly, Sebastopol.

Warf, B. and Sui, D. (2010). From GIS to neogeography: Ontological implications and theories of truth. *Annals of GIS*, 16(4), 197–209.

Wilson, M. and Graham, M. (2013). Neogeography and volunteered geographic information: A conversation with Michael Goodchild and Andrew Turner. *Environment and Planning A*, 45(1), 10–18.

Woodward, D. and Lewis, M. (1998). *The History of Cartography: Cartography in the Traditional African, American, Arctic, Australian, and Pacific Societies.* University of Chicago Press, Chicago.

5

The Cartographic Factory of Modern States

Bernard DEBARBIEUX
University of Geneva, Switzerland

5.1. Introduction

The academic literature on the use of cartography in state practice is extensive. It is mostly guided by the concern to account for two main research questions:

– What does cartography enable a state to know (e.g. about its territory, populations and resources) and do (e.g. guide and support state action in areas as varied as territorial defense, population control, public policy deployment or propaganda, etc.)? Works that address this type of question favor an instrumental approach to state cartography.

– How do states organize themselves to produce or use this instrument? We have here a wide range of analyses on the specialized services within the army or the administration, and the precise modalities for the production of cartographic coverage of state territories at different periods. The value of this work is of an operational nature.

This abundant literature, sometimes general (Branch 2014; Desbois 2015), and occasionally focused on specific states or periods (e.g. the collection of books: *The History of Cartography* 1987–2020), has rightly instilled and reinforced the widely accepted idea that cartography has been, since the origins of the modern

For a color version of all the figures in this chapter, see www.iste.co.uk/debarbieux/mapping.zip.

The Politics of Mapping,
coordinated by Bernard DEBARBIEUX and Irène HIRT. © ISTE Ltd 2022.

state, a central instrument and practice of its action. The approaches adopted to account for it are varied – sometimes descriptive, sometimes critical – and are guided by schools of thought that are just as varied (cultural studies, administrative geography, international relations, etc.).

However, from the 1980s onwards, another type of literature was added, more specialized, often even more theoretical, which can be capitalized on to problematize in depth the relationship between cartography and politics, in accordance with the general objective of this book (see the Introduction). These include works by geographers (e.g. Raffestin 1990; Pickles 2004; Farinelli 2009), historians (Akerman 1984, pp. 84–93), sociologists and philosophers (Lefebvre 1991 [1974], 2009 [1978]; Biggs 1999) and specialists in international relations (notably, Branch 2014), who outline or support the hypothesis that cartography is properly constitutive of the modern state from its origins around the 16th century, and that its evolutions have accompanied the transformations in the ways of thinking about the state and its missions in the centuries that have followed. From this perspective, much more than being a simple, convenient, or even efficient attribute of state policies, or an instrument of *ex post* legitimization of the state itself, cartography can be conceived of as a condition of possibility of this type of institution and of its policies from the modern period onwards. It is this line of analysis that this chapter wishes to contribute to in its own way. In doing so, it will not neglect the works mentioned in the Introduction; rather, it will seek to put them at the service of this thesis by paying tribute to the contribution of their authors to the empirical knowledge related to this theoretical approach.

5.2. The invention of spatial modernity in the Renaissance

It is now well established that the Renaissance invented a new culture of space in the West, and it is equally established that this new culture of space, specific to modernity, favored the sense of sight, even if the debate remains open as to the uniqueness or plurality of the "scopic regimes of modernity" (Jay 1988) invoked. The growing interest in cartography, since this turning point, was part of this new culture, but it was only one of its modalities. It accompanied the rediscovery of Euclidean geometry, which the English elites in particular were passionate about from the beginning of the 16th century, but also the pictorial revolution theorized by Leon Battista Alberti and emblematized by the use of the linear perspective (Edgerton 1975).

As for cartography itself, once emancipated from its explicit references to the cosmogony and theology characteristic of the European Middle Ages, it took on entirely new forms that exploited the resources of geometric measurement, guided by an ever-increasing concern for precision and accuracy in the representation of spatial forms and locations. But it also interacted with literature (Conley 1996; Klein 2001; Smith 2008) and painting: for example, with Leonardo da Vinci (Pinto 1976), in which the concern for objectivism was combined with that of an esthetic specific to this kind of production, or, later, in the Dutch paintings of the 17th century, which Alpers (1983) and Schama (1987) have shown to be guided by a cartographic vision of the world. In addition, the map entered private rooms and monumental galleries (e.g. the Map Room in the Vatican) as an ornamental element. Finally, maps were used in documents for a practical or legal purpose: the first land ownership plans (notably the first English estate maps) were intended to describe and secure property before the courts of justice, as well as contribute to this new culture of space, of which land appropriation was one of the modalities (e.g. Kain and Baigent 1992; Buisseret 1996); technicians and management bodies, such as those of collective water management in the Netherlands, also produced maps as early as the 15th century. In passing, it is not without interest to recall that the surveyor-mappers of the 16th and 17th centuries were sometimes also authors, such as Thomas Hobbes, whose philosophical contribution to this new culture of space was considerable.

Under these conditions, it is clear that the novelty and explosion of cartographic production in Western Europe in the 16th and 17th centuries were part of the artistic, scientific and social revolutions of that period. It is therefore possible to see it, along with other forms of spatial thought and expression of that time, as one of the dimensions of the social imagination of modernity. The use of this expression signals a borrowing from the propositions of Castoriadis (1987 [1975]) – a proper social imaginary is what institutes a specific society – and Taylor (2004) – on the proper and historical characteristics of the social imaginary of modernity – while continuing to extend them (Debarbieux 2019 [2015]), by showing that specific forms of spatiality are constitutive of these imaginaries. From the first half of the modern era, Euclidean geometry and its translation into cartographic production helped forge a social imaginary of space that made possible new forms of objectification and subjectivation (Conley 1996; Pickles 2004) the heart of the modern project and its appropriation by the political and economic elites of that time. The territorial state is itself the output of this cultural revolution.

5.3. State imaginaries of space and territory

The significance and various manifestations of what many authors have called the "cartographic consciousness" of the elites of the early modern period make it possible to relativize the inaugural character, long imagined as such in any case, of the very privileged relationship between state and cartography. In the 16th century, although cartographic projects were initiated by princes and states, they were not the only sponsors of this type of production, nor even the main ones, as in the Netherlands and England. On the other hand, it is clear that a very special relationship began to develop between modern cartographers and states, a relationship that has been described as "symbiotic" (Kagan and Schmidt 2007, p. 661), so much so that their respective practices were mutually conditioned.

The history of one of the most remarkable productions of the 16th century, the *Atlas of the Counties of England and Wales*, published in 1579 (see Figure 5.1), is revealing in this respect (Helgerson 1988, pp. 93–134; Barber 1992, pp. 57–98). The first edition of this atlas was the result of an exchange, initially without a formal commission, between Christopher Saxton, a map enthusiast and admirer of Ortelius' atlas for scholars and merchants, and Thomas Seckford, Master of the Requests of Elizabeth I. It was only gradually that the English crown took an interest in this work; the royal arms appear in later editions, as does the Queen herself in the frontispiece. Other contemporaries of this production quickly became aware of the political importance of this document. In 1592, a close associate of Elizabeth I, Sir Henry Lee, commissioned Marcus Gheeraerts the Younger to paint a portrait of the Queen, in order to secure her return to favor. The painting (see Figure 5.2) shows her standing on the map that opens Saxton's atlas. The visual device attests to two important and quite new things: on the one hand, the map is staged as a major attribute of monarchical power; on the other hand, it suggests that Elizabeth I can not only be conceived of as Queen of the English, but also as Queen of England; the map is there to suggest a symbolic equivalence between the monarch and her territory, while the standing position on the map evokes a form of territorial authority.

The example is quite singular and has no equivalent. Yet it appears to be paradigmatic. In a dozen years, an atlas born of a shared cartographic curiosity acquired the status of a state document, conducive to the administration of the kingdom – we will explain later how – and to monarchical staging. From then on, the fate of Saxton's atlas both participated in and attested to the adoption of a state imaginary of space that gave new meaning to the notion of territory. The expression, adapted from that of the social imaginary, suggests that the modern state, as it was invented in the 16th century between Spain and the United Provinces, via France and England, was henceforth mainly instituted by the relationship that bound it to its area of sovereignty and authority, and the use of the map was a full participant in this new imaginary. The transition between the destiny of an atlas, on the one hand, and the above statement, on the other, may seem quick and rash. It deserves to be put to the test with a corpus of practices and numerous documents.

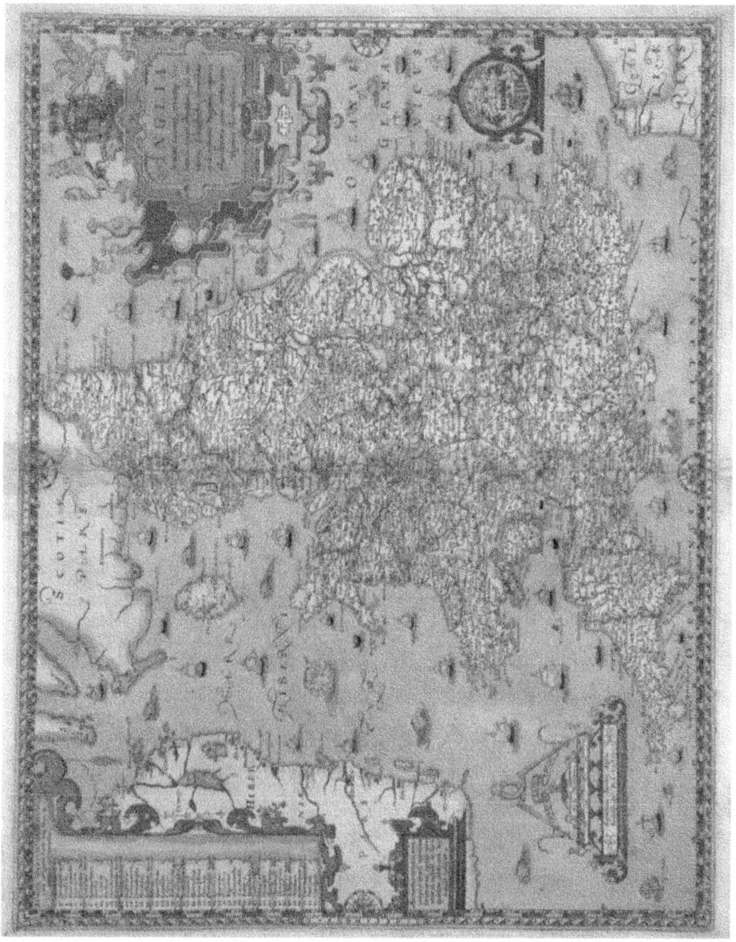

Figure 5.1. *Main map of Saxton's Atlas of the Counties of England and Wales 1579 (source: Wikicommons)*

Figure 5.2. *Portrait of Elizabeth I, Marcus Gheeraerts, ca. 1592. National Portrait Gallery, London (source: Wikicommons)*

The cartography that took hold around the 16th century accompanied a real revolution in the way political authority was conceived. In the Middle Ages, this authority was mainly exercised over individuals, according to the complex bonds of subordination between serfs and lords, vassals and suzerains. This form of authority does not lend itself easily to cartographic representation; it is more readily illustrated by lists of cities or fiefs, or even vignettes of urban landscapes accompanying official documents, all of which reflect these subordinate relationships. Conversely, modern cartography allows the reference space (that of Euclidean geometry) to be at the service of another form of authority – territorial authority – which is deployed over a particular geographical area. The 16th century was also the period in which the idea that sovereign authority must be exercised through the law was established through philosophical and legal treatises, such as those of Francis Bacon in Great Britain and Jean Bodin in France, or works of fiction such as *Utopia* by More (1516). But law is only ever valid in an area where authority is exercised and recognized. The law is applied in an isotropic mode, which is also that of territorial authority and Euclidean space. From then on, this common isotropy makes it possible to inscribe state authority in an order deemed natural, that of geometry, through the medium of the map, where coordinates, units of length and time, in short, measurement (Kula 1986 [1970]; Scott 1998) and the reference to the universal triumph then, supported in this by the map, the abstract space of the state authority "destroys the historical conditions that gave rise to it, its own (internal) differences, […] in order to impose an abstract homogeneity" (Lefebvre 1991 [1974], p. 370). Such is the imaginary of space that makes modern states become properly territorial (Herz 1957; Elden 2013) and differentiates them from empires and monotheistic churches whose universalist claims do not sit well with this political and legal vision.

It should be recalled that this territorialization of state authority coincided with the emergence of a new way of conceiving land ownership, which made the private owner a central figure in modern societies (Blomley 2003). The spatial frameworks for thinking about property and the territoriality of the rule of law have many affinities. As Schmitt (1950) and Arendt (1958) pointed out very early on, land appropriation and territorial appropriation are based on a similar exercise of spatial partition, the trace of which can be found in etymology: the term *nomos*, which designates law in Latin, has the idea of sharing an area (initially pastureland) between owners at its roots. In both cases, therefore, there is partition, and in both cases the use of geometry and cartography is a guarantee of accuracy. This is what encouraged Schmitt to say that the law is the "constitutive act of the spatial order" (Schmitt 2001 [1950], p. 75).

Moreover, at the turn of the modern era, the state acquired the status of an actor in its own right, with its own identity, distinct from that of the monarchs. For the latter, it took time to establish the legal basis for a clear distinction between the area over which they were able to deploy their authority and their personal domain. This depersonalization of the state, clearly attested in the 17th century, making it all the more useful to have recourse to a set of symbols that give it flesh and "aggregate the feelings of loyalty and trust" of its subjects (Walzer 1967). The map of the state's territory is one of these symbols, and even one of the most used, thanks in particular to the increased dissemination made possible by the widespread use of printing from the end of the 15th century.

It is for this set of reasons that we can say, contrary to a well-known formula, that with the emergence of this state imaginary of space, the map was the territory, or, to put it in a more scholarly way, that it performed the territory or participated in its ontogenesis; it was both the privileged figure and the condition of possibility, just as the territory became both the matrix and the condition of possibility of the modern state. However, this revolution obviously did not take place in one place, at one time, and around a single document (e.g. Saxton's atlas), however paradigmatic it may be of the revolution in question. It spanned several centuries and was manifested in several parts of Western Europe. A few milestones can be marked.

Princely commissions for maps covering an entire kingdom or duchy, possibly supplemented by detailed maps, only became frequent in the second third of the 16th century: a map of Brabant commissioned from Jacob van Deventer; a map of Bavaria that was due to Philipp Apian; a set of 20 maps collected in the "Escorial Atlas" (1530s–1540s) depicting more than 8,000 towns and cities of the Iberian peninsula, composed at the request of Charles V and extended by a complete map in the 1550s commissioned, in turn, by Philip II (see Figure 5.3). However, this type of map did not quickly become the dominant form of territorial description. Philip II, who had been trained in map reading by his father and had been a great enthusiast (Buisseret 2007, pp. 1069–1094), also commissioned surveys for various parts of his empire (Mundy 1996) and a series of city portraits reminiscent of late medieval iconography. This movement, which began in the western part of Europe, spread very gradually throughout Europe in the following centuries and reached Russia in the 18th century, with the *Atlas Russicus* (1745) commissioned by Peter the Great (Bagrow and Castner 1975). In its wake, many states set up specialized military or scientific services, sometimes attached to the monarch, to ensure the continuity of projects, the conservation of documents and the training of agents (Biggs 1999).

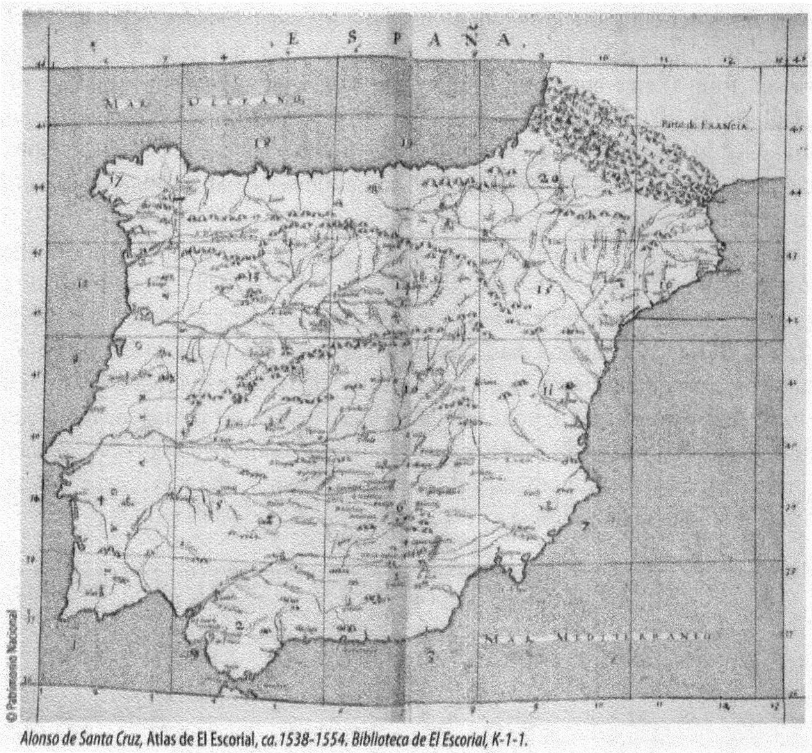

Figure 5.3. *General map of Spain and Portugal, from the "Atlas Escorial", by Alonso de Santa Cruz (source: Wikicommons)*

At the same time, the state's cartographic production also increased, multiplying the types of documents and intensifying and diversifying the forms of territorialization of the state. The reign of Louis XIV (1643–1715) is exemplary in this respect. Although the kings of France were very early to produce maps of their kingdom (Pelletier 2007, pp. 1480–1503) by hiring the services of mathematical cartographers (starting with Oronce Fine for the map of 1525), and although a strong interest in maps led Henry IV to commission the first atlas of France (completed in 1594), it was with Louis XIV that the attention paid to cartography reached its peak. During the first 30 years of his reign, monarchical symbolism and mythological (Apollo, Jupiter) and cosmogonic (the Sun) references took precedence; but from the 1680s onwards, it became more down-to-earth. The monarch hired the services of the *Académie Royale des Sciences* to produce a map of the kingdom using the most advanced techniques of the time. The initial project by Jean Picard was to base the map on a meridian line – the "*grande traverse*" linking Dunkirk to Perpignan – and another line encircling the kingdom along the borders and coasts (Konvitz 1987). It was, therefore, a question of basing the representation of the territory on a backbone borrowed from the geometric grid of the earth and on a clear identification of the external limits of the territory. The map that resulted soon afterwards gained enormously in accuracy; but by considerably reducing the area of the country, it made Louis XIV say, with unfeigned disappointment, that his kingdom had lost more in one map than his armies had conquered (Konvitz 1987; Petto 2007). It is this map that formed the basis of the great project, which covered more than a century, desired by Colbert from the 1660s: the detailed coverage of France entrusted to several generations of astronomers and geometers of the Cassini family (Pelletier 2002). But before that, Vauban had been commissioned to ensure the defense of this same territory. In addition to the detailed maps that allowed him to optimize the location and layout of the military works he designed, he added an overall vision that was itself highly cartographic: the contribution of each element to the protection of the land borders and the optimal shape – the "*pré-carré*" – that the territory should take to optimize its defense.

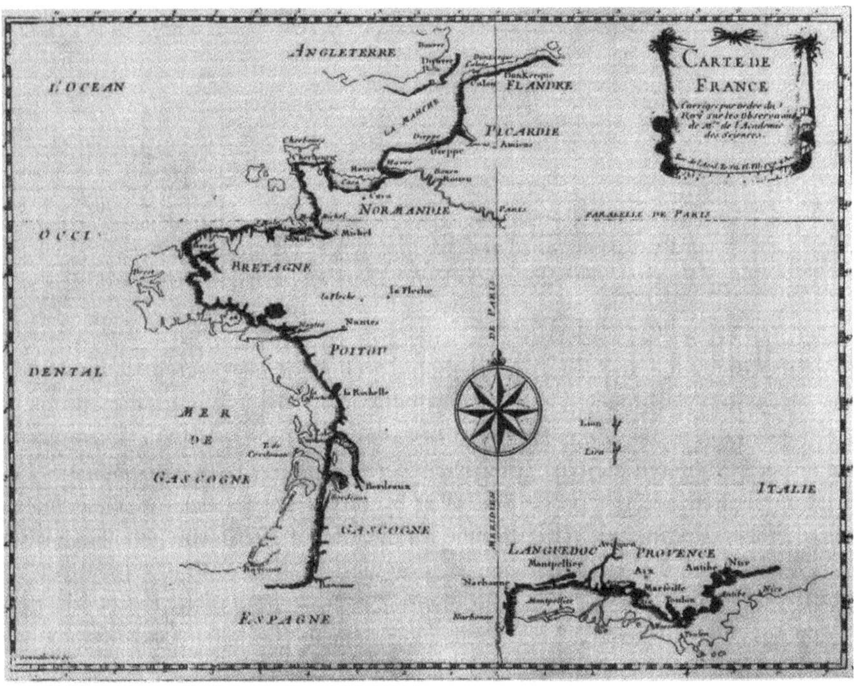

Figure 5.4. *"Map of France, corrected following the order of the King on observations made by the Académie des Sciences", 1682, Mémoires de l'Académie royale des Sciences, T7 (source: Wikicommons)*

Another manifestation of the institution and triumph of territorial states is in international relations. Very early on, maps were attached to the negotiation of inter-state treaties. Tuscany was early in this respect. However, the model of the territorial state was slow to take hold in peace treaties, even in the 17th century. The treaties of Westphalia (1648), in which some authors readily see the irruption of this model in international relations, do not deserve this honor for others (Osiander 2001; Branch 2014, pp. 125–128): no maps, no descriptions, nor properly territorial demarcations, but an abundant use of lists of cities and places that established the relationships of authority that bound them to the princes of the time. The Treaty of the Pyrenees (1659), which stabilized the Franco-Spanish border for centuries, marked a change in direction; the general delimitation of French and Spanish sovereignty was not achieved with the help of maps: the principle of "*eaux pendantes*" (the watersheds of the rivers flowing north and south) was mainly relied upon, in other words, a different geographical reading grid. Only the Supplementary Treaty of Llivia (1660), which sealed the partition of Cerdanya, was given a map, although Peter Sahlins dates the first efforts to linearize the border to the 1730s, at the initiative of Spain, which then set up a system of taxation on land ownership (Sahlins 1989, pp. 85–86). However, at the same time, the sovereigns showed how much symbolic value they attached to the precise delimitations of their kingdom: the meeting of Philip II and Louis XIV that preceded the marriage of the French king to the Spanish infanta (1660) was organized on an island in the Bidasoa River, divided into two watertight camps; the scene of the meeting, immortalized by the French painter, Charles Le Brun, even shows that the arrangement of the place – the organization of the room, the arrangement of the carpets, the routing of the two processions – celebrates the exact line that was supposed to serve as the border between the two countries (Debarbieux 2017, pp. 117–136).

Despite the survival of pre-modern customs, many signs accumulated in the 17th century, showing the rise of the territorial prism through which international relations are conceived. However, many authors agree that it was at the end of the 18th century, particularly in the context of the carving up of Poland (Lukowski 1999), which gave rise to several cartographic allegories (see Figure 5.5), and even more so at the end of the Napoleonic Wars, that the model of territorial states became fully established in international relations (Branch 2014).

The territorial state, established by a new paradigm of authority, the fundamental role conferred on the law and the adoption of a new conception of space, thus gradually took shape between the 16th and early 19th centuries. England, France and Spain played a decisive role in its configuration, before it became a model for other European states. A whole range of cartographic production from this period – state maps, regional maps, detailed maps – was part of this state institution. But in the second half of this period, another parameter already contributed to the invention of this particular form of state, the nation-state.

Figure 5.5. *"The Cake of Kings", engraving by Noël Lemire. This allegory of the First Partition of Poland (1772) depicts the King of Poland, Stanislaw Poniatowski, struggling to keep his crown on his head and resist the division of his territory (actually his map) coveted by Catherine of Russia, Frederick II of Prussia and Emperor Joseph II (source: Wikicommons)*

5.4. The construction of the nation-state through the map

Indeed, another modality of state institution through cartography, which barely post-dates the appearance of territorial states, and which is destined to take on central importance, arises from the identification of the nation as a collective of subjects, and then of citizens, coextensive with the state territory and the basis of the state's legitimacy. This conception of the nation runs through the works of political philosophy from the 17th century onwards and Thomas Hobbes' inaugural book, *Leviathan* (1651) – and its central idea of a population instituted by social contract giving power to the prince – and flourishes in Rousseau's *The Social Contract* (1762), which both extends it and criticizes it in the most radical way. The latter imposed the idea, destined to a revolutionary destiny, of national sovereignty; but if, with Rousseau, sovereignty passed from the hands of the prince to those of the nation, it remained territorial, that is to say, justified by coexistence in a common area of identification, over which the nation had a legitimate claim.

For their part, historians have suggested that national sentiment, initially more cultural than political, pre-exists these theories and that cartography is no stranger to its emergence (e.g. Wintle 2016, pp. 271–287). For Helgerson (1992), it was already at work in England at the time of Saxton's atlas; according to him, it was a complex of texts, those of William Shakespeare or Philip Sidney in particular, and images that participated in a "spatial visualization of England" of which the cartographic point of view was the central character. In this context, Saxton's atlas is as much a vehicle for the monarchy's assertion of authority over a now insular territory as it is a moment when we see "the sense of both local and national identity at the expense of an identity based on dynastic loyalty" (Helgerson 1992, p. 114). The popularity of the reference to *Britannia*, a name given to both the territory and the British people, notably by William Camden (from 1586 onwards) on the one hand, and to a figure personifying the nation on the other hand, contributes to this iconographic and ideological assimilation of the nation and its territory.

The demands for collective emancipation from a distant monarchical authority also provided a context for national assertion through the invocation of a patriotism that was also readily mediated by maps. A well-known example is the succession of maps sometimes named *Leo Belgicum*, and sometimes named *Leo Hollandicus*, which cover the six decades (between 1579 and 1648) that led to the secession of the Spanish provinces of Northern Europe (van den Heijden 1990; Ricci 2015). The first ones propose a personification of the territory of the then so-called "Belgian" provinces of Philip II's Empire, by taking up the symbol of the lion, frequent in the arms of these provinces. The aim was neither strictly national nor secessionist, as the association of these provinces with the image of the animal was well known to all, and had already been used by Charles V himself. But later editions gradually took on a nationalist and secessionist tone as the independence of the northern provinces approached, justifying *Leo* becoming *Hollandicus*. While the personification varied little over the period, the vignettes that accompany the map changed quite dramatically. After having favored the arms of the provinces on either side of the map, these side illustrations turned to effigies of rulers (1598), then figures in popular costume (1617) and topographical portraits of prosperous cities (1611, 1648). The 1648 map (see Figure 5.6) by Nicolaes Ioannis Visscher was published the same year that the Treaties of Westphalia were signed, sanctioning the independence of the United Provinces. On the sword carried by the lion, the phrase *patria defensio* appears, praising the patriotic impulse that led to this independence. The multiplication of these personified maps until late in the 18th century and their popularity make it possible to see them as an early expression of national pride in cartography, as well as a vehicle for the emergence of one of these European nations.

Figure 5.6. *Leo Hollandicus (Comitatus Hollandiæ denuo forma Leonis), by Claes Jansz Visscher, 1648. Universiteits Bibliotheek van Amsterdam (source: Wikicommons)*

Another modality of the cartographic institution of the nation lies in the systematic campaigns of surveying and land planning when they are driven by a genuine national ideology. The contemporary cartography of the early United States is instructive in this regard. The first post-independence printed map, by Abel Buell in 1784 (see Figure 5.7), combines a statement of fact – an aggregate of 13 colonies arrayed against England, clearly differentiated from the Canadian provinces – with a spatial project – a claim to expansion between Appalachia and Mississippi. Buell's map circumscribes a young nation and gives it a territorial horizon.

But this is only one dimension, one scale really, of the American national project. The other lies at the heart of the constitutional project and the vision of society that it embodies. Strongly influenced by 17th-century philosophers such as John Locke and James Harrington, who theorized about the role of land ownership in the emergence of new forms of statehood, the first steps of the American republic were guided by an egalitarian conception of a society of smallholders. Thomas Jefferson, one of the founding fathers and third president of the United States (1801–1804), was the main advocate of this vision. He was one of the many who planned the creation of new federal states beyond the Appalachians (see Figure 5.8); it was also him who suggested a policy of land allotment (but also of their capture to the detriment of the Native American populations) that was intended to be egalitarian and democratic (Land Ordinance of 1785). Both operations were then aligned with the grid of geometric coordinates that made it possible to project from afar, in a universalist and rationalist mode, a cartographic vision of the occupation of the territory that was common to all of its scales (Pattison 1957; Blomley 2003; Maumi 2007).

Although Jefferson's initial project underwent numerous adjustments and the political philosophy that animated it was successfully challenged by others, the cartographic project became imprinted on the landscape (fields, roads, urban clusters of houses, etc.) and in people's minds as an indelible signature of the American nation and territory. From the Renaissance to the middle of the 18th century, the geometric coordinate grid borrowed from Ptolemy had the function of objectifying the territorial extent, specifying the location of its components and, in passing, relating any map to a reference point presented as external and natural; with Jefferson and his contemporaries, it became the cognitive and practical instrument of the appropriation of an area that was to become a state territory and a national project.

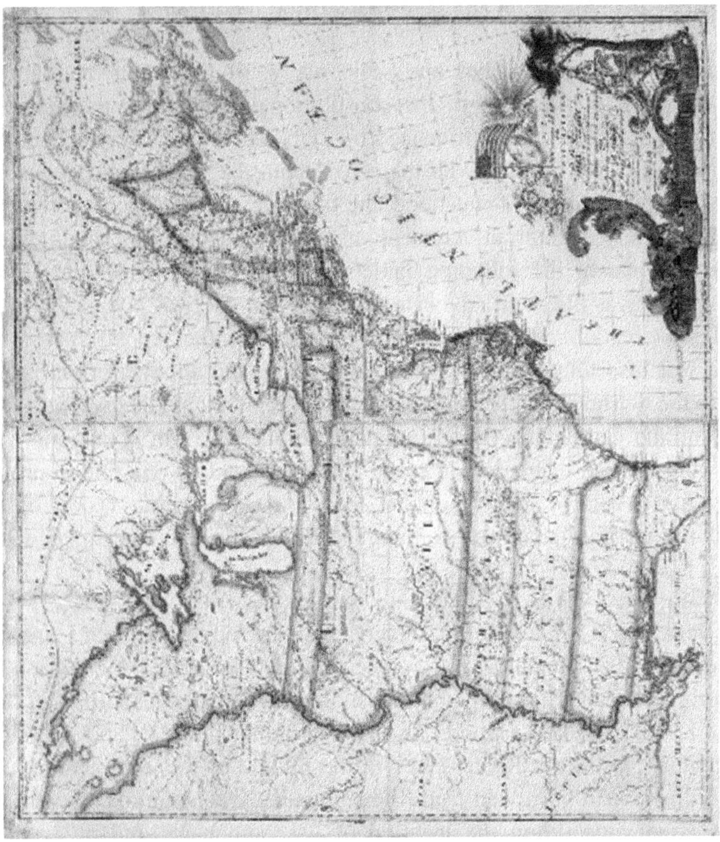

Figure 5.7. *Abel Buell. A New and Correct Map of the United States of North America: Layd down from the Latest Observations and Best Authorities Agreeable to the Peace of 1783. New Haven, Connecticut. 1784. Library of Congress (source: Wikicommons)*

Figure 5.8. *A map of the United States east of the Mississippi River, in which the land ceded by the Treaty of Paris is divided by parallels of latitude and longitude into 14 new states. By David Hartley and Thomas Jefferson, 1784 (source: William L. Clements Library, University of Michigan)*

Another type of nation-state institution through cartography corresponds to states created through a process of identification and differentiation of linguistic communities that became political nations. Illustrations of this type can be found on the rise, evident from the 19th century onwards, of the reference to language in the national imagination. Starting with Germany (Labbé 2010, pp. 49–72; Hansen 2015), formulated by philosophers at the beginning of the century, Johann Gottfried von Herder and Johann Gottlieb Fichte in particular, the idea of unifying German speakers became a popular theme of political discourse in the first half of the 19th century. The debate was illuminated by an abundant production of maps, which, while reflecting Germanic nuances and dialects, clearly delineated an area, inviting the coincidence of the territory of the nation and the territory of the state. The maps by Karl Bernhardi and Heinrich Berghaus are two such maps. The first (see Figure 5.9) was displayed in the church where the Frankfurt Diet met in 1848, which brought together liberal and nationalist politicians who were angry that the Treaty of Vienna had restored the old regime and perpetuated the territorial fragmentation of the Germanic area. Bernhardi himself was a member of the Council as an elected representative of Prussia. These two maps were arguments for the creation of a German state based on a linguistically defined nation.

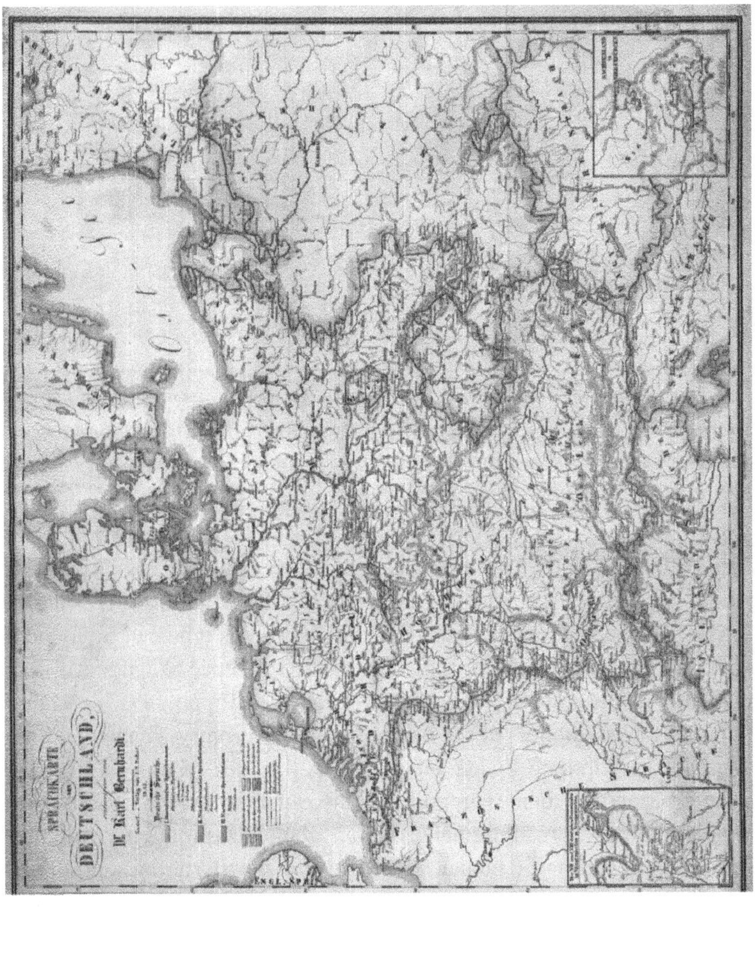

Figure 5.9. *Karl Bernhardi, Sprachkarte von Deutschland, Kassel, 1843 (source: Wikicommons)*

The diplomatic negotiations of the Paris Peace Conference in 1919, which followed the First World War, also gave considerable space to linguistic maps (Ter Minassian 1997; Boulineau 2008). In the spirit of the "right of self-determination" promoted by the President of the United States, Woodrow Wilson, during the elaboration of the new borders of Central Europe, such maps were produced and expert reports were requested from linguists and geographers to help diplomats in the carving up of the Austro-Hungarian Empire. Although the territories created or re-drawn, and the boundaries chosen in the Balkans and the Danube basin were ultimately determined by the power relations between the parties involved and other factors, these maps played an important role and language became a central element of state policy in the region. A similar return to linguistic and religious maps guided the Dayton negotiations (1995) that led to the breaking up of Bosnia (Johnson 1999).

The importance of general maps in the institution of modern states in the 16th and 17th centuries was therefore matched by the importance of various kinds of maps in the institution of nation-states between the 17th and 20th centuries. The latter has been noted on several occasions: by Jacob, who writes that they constituted "the visual cement of the feeling of national identity" (1992, p. 86) and that the mastery of space to which they contribute "undoubtedly constitutes an essential stage in the process of acculturation of the individual, as well as in the formation of a national consciousness and identity" (1992, p. 310); by Cubitt who writes that they "have the power to transform discourses of national identity" (1998, pp. 128–129); or by Herb and Kaplan who believe that "only maps are able to communicate a precise image of the limits of a nation and foster a territorial consensus" (1999, pp. 23–24). A similar demonstration has been conducted on the influence of the so-called Dufour map on Swiss identity, at the very moment when Switzerland was adopting a modern, federal constitution (Gugerli and Speich 2002).

Certainly, this second generation of instituting maps owes much to the fact that they participate in the birth or politicization of national imaginaries that give the idea of state territory a new meaning: an area of settlement and homogeneous culture and a space of identification. In doing so, they do not invalidate the achievements of the previous period, which gave precedence to the institution of the state through the adherence to territorial sovereignty via the image. It enriches it with a new contribution and paves the way for a reinvention of the relationship between state and individuals; if the latter, patriots or citizens, construct new territorial identities and compose new political bodies through the cognitive and affective mastery of these territories, they are also affected by state policies which, within the same territorial framework and with the help of *ad hoc* maps, contribute to adjusting state territory and national territory (Agnew 1998), and to institutionalizing new social relations within the nation-state (Tilly 1975; Brenner and Elden 2009).

5.5. The colonial adventure: a counter-illustration?

The colonial experience is, to some extent, a counter-illustration of many of the lessons of the previous developments. Note that cartography did not have an important place in this adventure; it was ubiquitous and multifaceted, especially in exploration, in the division of territories between colonial powers, in the assertion of the sovereignty of colonial powers and in the field descriptions that Carter has said were part of a " strategy for translating space into a conceivable object, an object that the mind could possess long before the lowing herds" (Carter 1988, p. 329). Also note that cartography played a secondary role in the idea of empire and their construction: it was decisive to the point that one could write: "the empire exists because it can be mapped" (Edney 1997, p. 2). But this imperial cartography was never intended to help establish colonializing states[1].

Furthermore, colonial administrations were never concerned with delineating populations or nations, with or without supporting maps, with their corresponding territories, in order to create complementary entities – state, nation, territory – on the model of modern states. Colonial cartography aimed, almost exclusively, to differentiate and locate Indigenous populations, in order to optimize their management or control. From then on, and in contrast to these examples, it was mainly the objectives of conquest and prestige linked to this conquest, and of control and management of resources and populations that prevailed in the intense cartographic production in the colonial context.

The example of British India is illuminating in this regard. The conquest was supported by the production of maps reporting on expeditions and detailed surveying to retain and fortify borders. The model that was aimed for was that of "scientific boundaries", in the words of British Prime Minister Benjamin Disraeli; in other words, one that consisted of identifying, through the science of military strategy and meticulous cartography, the boundaries deemed most likely to

1 Even though the cartographic services set up to support Spanish and Portuguese colonization – such as *La Casa da Mina* in Lisbon and the *Casa de la Contratación* in Seville – may have had a subsequent impact on the production of maps in the Iberian Peninsula (see Buisseret 2007, pp. 1069–1094); even if certain devices invented in the colonies – such as the projection of land and administrative boundaries from geometric markers even before the places had been explored – could later be taken up in the territorial states, as we have just seen for the United States.

guarantee control of the margins of the Empire, even if it meant combining imported and local know-how (Raj 1997). In addition, there was also a cartography that aimed for a wider knowledge of the Indian Empire and its constituents: first, an overall cartography which between the end of the 18th century and the beginning of the 20th century (Edney 1997) moved from a representation of the complexity of territorialities inherited from the Mughal Empire and the divisions of the colonial administration (notably the maps of Hindoostan that James Rennell published in the 1780s) to a simplified map asserting sole imperial authority over a homogenized area; then a more detailed and systematic cartography carried out under the responsibility of the Surveyor General of India from 1814 onwards; finally, an administrative and legal cartography that specified the modalities of the colonial regimes retained; this cartography details, in particular, the status of buffer states (Kashmir, Nepal), that of the zones conceded to hill tribes, such as the Naga or the Chin, officially subjugated but which had *de facto* autonomy, and that of the zones that were fully subjugated (see Figure 5.10).

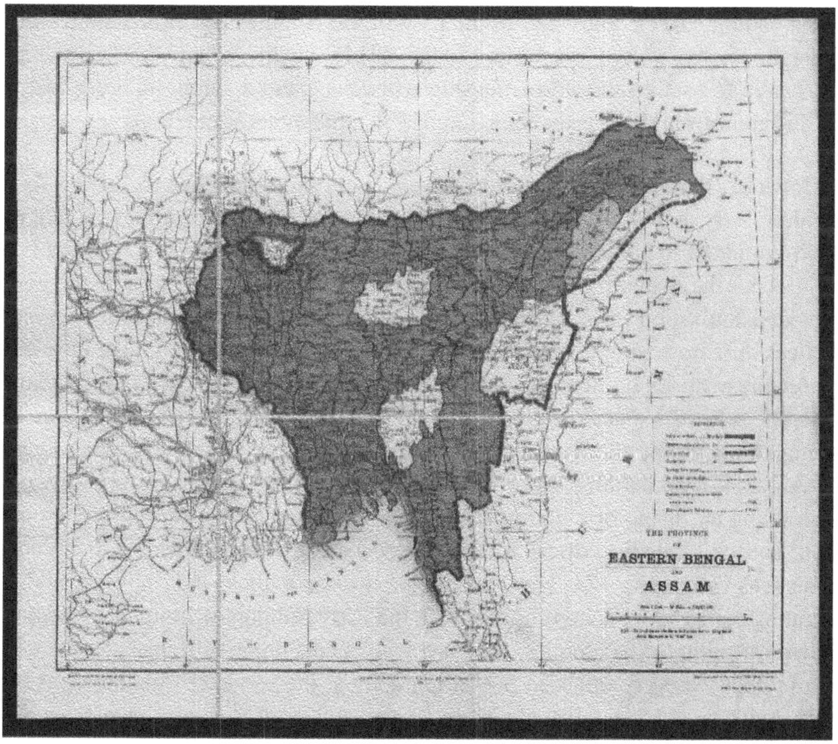

Figure 5.10. *The Province of Eastern Bengal and Assam, 1908, Survey of India (source: The British Library)*

This type of map reflects, through the range of colors used, the diversity of administrative management methods in the margins of British India. The external limits are also represented in a variable way: an outer line which marks the extension of territorial claims toward Tibet or the contact with Burma, another British colony; and an inner line, at the contact between rice-growing plains and mountains, which distinguishes the variable intensity of the colonial hold. An administrator of the Viceroyalty distinguished these two lines in 1910: "Up to the Inner Line we administer in the ordinary way. Between the Inner Line and the Outer Line we only administer politically" (Lord Hardinge, quoted by Debarbieux and Rudaz (2015, p. 170)).

As illuminating as it is, the Indian example does not apply to all colonies, not even to all of those of the British Empire, nor to maps that feature the entirety of that empire in a didactic or propagandistic mode (Avila 2015, pp. 191–204). In the case of the French Empire (Laboulais-Lesage 2004; Thomas 2004), maps identifying resources and populations and those facilitating the conquest and control of the corresponding spaces dominate. In the case of the colonies and protectorates of the Maghreb, many works (notably, Blais 2014) have shown that cartography contributed to establishing a relationship with space and a symbolic order that was very different from those prevailing among the Indigenous populations.

However, the situation changed radically with the independence of the former colonies. Very early on in their independence, most of the American colonies quickly imitated the practices that were then prevailing in Europe (Dym and Offen 2011). After the Second World War, the new African and Asian states that emerged from decolonization took up the instituting potential of certain cartographic practices, foremost among which was the full-frame representation of the territory of sovereignty with its borders. In the Maghreb, they prolonged, in a barely adjusted mode, the practices that prevailed in the metropolises, such as detailed topographic coverage; but unlike the French colonial administration, which was committed to differentiating populations, these states willingly put cartography at the service of constructing an image of a homogeneous nation by neglecting the representation of ethnic and cultural diversities. A similar process has been observed in Indonesia (Anderson 1983, pp. 174–178). In India, the state took the indigenization of instituting cartography a step further; with the early training of Indigenous people in modern cartographic techniques, the state developed a highly efficient service within its army that was very quick to put cartography at the service of the nation (Cohen 1971). Independent India then claimed to exercise full sovereignty to the exact limits of its territory. It then renounced the existing statutes of autonomy on the margins of the Empire and thus removed the presence of Indigenous populations from the map, to whom the Empire had assigned a special status, for example, the Nagas on the borders of Assam and Burma (Franke 2005). Cartography thus contributed to the negation of internal tensions and to the building of the Indian nation: "the anxiety surrounding questions of national identity and survival is particularly acute [...] the physical map of India gains ubiquity as an iconic representation of the body politic" (Krishna 1994, pp. 508–510).

However, while the colonial legacy may have facilitated, or even encouraged, the adoption of the territorial state model and its corresponding cartography by the newly independent states, this influence has also played out in states that have escaped direct colonial control. Indeed, it has been shown that Persia, Afghanistan (Hopkins 2007) and Siam (Thailand) (Winichakul 1994) "indigenized" Western norms of statehood for their own purposes and proceeded to "territorialize their political identity" (Hopkins 2007), notably by demarcating their borders using processes borrowed from modern cartography.

5.6. Maps and the government of things and people

So far, this chapter has focused on the ability of modern states and nation-states to construct themselves as such on the basis of the area of authority and law enforcement that they control and display through maps. But beyond this political and legal function, we have come across other state uses of maps: cartography for military purposes, cartography of populations differentiated according to cultural characteristics, cartography of soil and subsoil resources, etc. While these productions are not always independent of the first mentioned, they do not all contribute to the same degree to the institutionalization of these states. Some are certainly more instrumental; in other words, guided by management objectives, than by the assertion of authority and legitimacy. In order to distinguish between them, we can look at a few types of production over the long term.

Administrative and political maps on a sub-state scale, which enable the designation of the constituent entities of a country, have been omnipresent throughout the modern era. They define the areas over which the constitutional, legal and regulatory powers of local and regional authorities, more or less autonomous, are exercised. They are thus part of the institutional relations that bind a state and the different parts that its territory is composed of. While they initially highlight the diversity of the status of these entities, often born of the conditions of their attachment to a state territory, these maps tend to place the representation of these entities at the service of that of the state itself. Thus, Saxton's atlas opened with a general map of England and Wales, but then reviewed the various counties under royal authority. These county maps are very similar, except for a variable scale; by using recurrent graphic codes, these maps participate in the idea, promoted by the English monarchy, of an equivalence of the parts and their subordination to the state territory. The atlas that Philipp Apian made on commission of the Duke of Bavaria in 1568 – the Bairische Landtafeln – presents this same characteristic (Buisseret 2007, pp. 1069–1094)[2].

But there is another case, even more edifying in this respect. These are state maps that aim to establish a political and administrative division that is itself guided by this concern to harmonize and standardize the constituent parts according to a state project. Here, we find the example of Jefferson's map and the planning of the creation of the federal states located west of Appalachia. Also included in this category is the departmentalization of France at the beginning of the French Revolution: the initial project, which we owe to Robert de Hesseln, planned to divide France into squares, also based on parallels and meridians, by virtue of a similar belief in the virtues of geometry and the reference to a so-called natural order. Although this principle of division was not strictly adhered to in the end, the departmental territories that were retained, which were similar in size, reflected the desire to standardize the administrative partition of France and, with it, the country's intermediary administration organization (Ozouf-Marignier 1989; Biggs 1999). This exercise in standardization did not go unnoticed by the contemptuous critics of the French Revolution, such as Edmund Burke: "I cannot conceive how any man can... consider his country as nothing but carte blanche – upon which he may scribble whatever he pleases" (Burke, cited in Biggs (1990, p. 330)).

The administration of the state territory through maps has progressively taken on considerable importance since the 19th century. It is part of a logic of population and resource management that Foucault (2004) has described as governmentality (Hannah 2000), largely independent of the logic of authority. The relationship between the state and territory deserves to be analyzed here, less in a substantive mode than in a procedural one, in other words, from the point of view of the concrete actions that the state is led to carry out. From this perspective, the map, along with other instruments such as statistics, to which it has become closely linked, can be conceived of as a "technical device with a generic vocation that conveys a concrete conception of the political/society relationship and is supported by a conception of regulation" (Lascoumes 2007, p. 3). It also fully participates in

2 The Atlas of El Escorial, on the other hand, is made up of plates that do not aim to show the constituent entities of the Hispanic lands of the empire of Philip II/Charles V, but rather the equivalence of the places that constitute this same territory. Much later, the detailed topographic maps that many states adopted after Cassini's map also adopted, but with many details and refinements, such uniform codes reflecting the same concern for control.

the idea that "space becomes a 'project' of the state through the mediation of the administration" (Alliès 1980, p. 10).

One might think that this state concern with administering resources through maps is as old as modern cartography. After all, Renaissance Italy – which has been called "by far the most cartographically conscious part of Europe" (Harvey 1980, p. 58) – gave rise to maps attentive to the details of the territory controlled by Florence, Venice or the Pope in a concern for knowledge and intervention; but what is aimed at then is sometimes, as with Leonardo da Vinci or Cristoforo Sabbadino in the Veneto region (Casti 2007, pp. 874–908), the control of waterways for military and supply purposes, therefore in the manner of an engineer, and sometimes the control of resources, as with Machiavelli, but this is not a concern for administrative management as such. Foucault suggests that governmentality, as such, is a process of rationalization and technicization, relying both on a series of specific apparatuses of government, and a body of knowledge. From then on, the priority in the use of the map, as both knowledge and an instrument, was no longer conquest and control, but the optimal organization of a territory and a population, and the development of wealth.

From this perspective, the statistical and thematic map has played a dominant role since the 19th century. This role is, in a way, the equivalent, in terms of governmentality, of the role played by the topographic map in controlling the territory itself. The "*Carte figurative de l'instruction populaire de la France*" (see Figure 5.11), designed in 1826 by Charles Dupin, a French engineer, scientist and politician, is considered to be the first map of this type. Palsky (1996) notes that Dupin's map was designed well after the first quantitative diagrams appeared in the 17th century: "this discrepancy is understandable if one considers the state of statistics in the first decades of the 19th century, from the point of view of both its sources and its methods. Statistical mapping relies on homogeneous, centralized and published sources" (Palsky 1996, p. 59). Such sources could only be provided at the time by state statistical institutions (offices), which only really developed from the 1830s onwards "as one of the indispensable attributes of nation-states in the process of being constituted or aspiring to be so" (Desrosières 2000, p. 123).

Another particularly illuminating illustration of the logic of governing by maps was provided by Didier (2007). He studied the implementation of a federal survey on the state of the country's farms during the New Deal era, and the use of a cartographic survey method. Didier shows that this exercise cannot be understood as an instrument of control or "discipline" as Foucault understood it, but as a practice that both produces knowledge and reconfigures those who request it, in other words, a component of governmentality. And Didier (2007) concludes: "Governing by maps is not only producing maps, it is also producing their administration" (Didier 2007, p. 64).

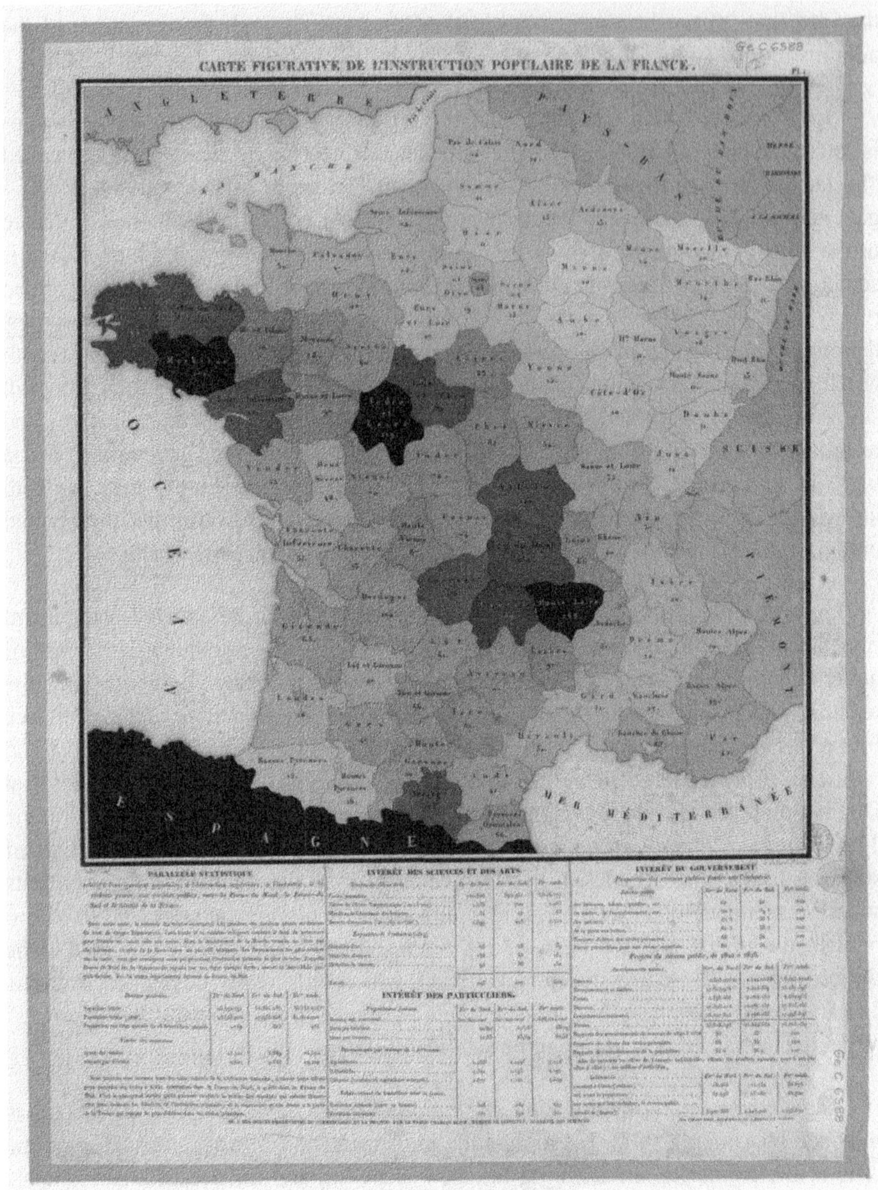

Figure 5.11. *Figurative map of popular education in France, Charles Dupin, 1826. Bibliothèque Nationale de France (source: BNF Gallica)*

It is also with the rise of this concern for governmentality that state policies for spatial planning emerge, which we know are richly endowed with maps. These maps may have an operational purpose, in particular when they draw a general balance in the location of activities and infrastructures, but they may also have a symbolic purpose: to provide an image of the concern for the public interest, which the state in question can claim. In both cases, they make the state's territorial action possible and remind us that the state is fundamentally territorial in nature. Chapter 9 of this book explores what happened to DATAR, the French administration in charge of this mission from the 1960s to 2010.

The emblematic value of the map, including in land use planning, reminds us in some ways of the use of maps in Elizabethan England. But it is also fair to point out that it has been particularly common since the 19th century. Whether in the form of a map displayed at the back of a classroom, propaganda maps, schematic maps (e.g. the French "hexagon") or logo maps – "map-as-logo" (Anderson 1983, p. 175) – reproduced endlessly on stamps or letterheads, these signature maps are part of the communicative paraphernalia aimed at reminding us how much the state exists only through its territory and how much it is appropriate to cultivate the nation's attachment to it.

Maps with a military vocation are of a different type. Very old and numerous since the 15th century, they bring us back to a logic of control and command of the territory. Machiavelli and Guicciardini recommended that every state should have one for the progress of its armies in foreign territory. Before them, Charles VIII had ordered a general map of the Italian peninsula before embarking on his transalpine expedition at the end of the 15th century (Buisseret 1992; Pelletier 2007). Later, Vauban, as we have seen, made it a preferential tool for the planning of military works, and used it to outline the margins of French territory. And this very specific type of cartography continues today (Williams 2011). All together, these state commissions contributed greatly to the technical progress of cartography (Hale 2007, pp. 719–737) and to the familiarization of princes with this type of document. But at this stage, they were not involved in the administration of resources and populations. On the other hand, from the 16th century onwards, princely orders began to combine military, economic and administrative objectives. This was the case during the reign of Henry IV of France, when, on the advice of Sully, the kingdom was equipped with a large number of local maps intended for jurists, agronomists and the army (Buisseret 1964). In the 18th and 19th centuries, with the systematic and detailed mapping of territories (the Carte d'État-Major in France, the maps of the Board of Ordnance in the United Kingdom), the military objectives of cartography tended to merge in part with those of resource and infrastructure management.

5.7. Current state of the issue

The hypothesis of this chapter is that cartography has played a decisive role in the establishment of modern states, rather than accompanying their deployment, and in the design of public policies, rather than supporting them. However, it deserves to be assessed in light of the considerable changes observed over the past few decades in the production and use of maps, and in the evolution of the very conception of the state and its territoriality (Agnew 2009). These changes are numerous and abundantly detailed in this book. We will limit ourselves here to listing those that question the symbiotic relationship that has been built up between the territorial state and the modes of cartographic visualization of its territory.

The very special relationship between territorial sovereignty and cartographic production has been undermined in recent decades, to the extent that it has been said that we have moved on to a "post-mapping sovereignty" (see Chapter 3 of this book). This special relationship was the founding principle of the modern state in Europe, and was intensified in the 19th and 20th centuries by the growing influence of states in the cartographic enterprise. Monopolistic claims to detailed cartographic production, sometimes under the aegis of the military, coupled with a culture of secrecy, as in the well-documented case of the USSR, were the main illustration. However, the advent of aerial espionage missions, which were admittedly illegal from the outset, and then the proliferation of satellite images of the earth's surface, which were not regulated by international law, deprived states of their exclusivity or of the regulation of the production and dissemination of images of their territory, some of which were very detailed. The various protests or attempts to ban them by many states have proven ineffective on the whole (Desbois 2015, p. 105 onwards).

The production of this terrestrial imagery, freed from the framework of territorial sovereignty, was initially the prerogative of a few states, notably the United States through NASA. However, since the 1990s, the production of this imagery has largely passed into the hands of private companies, the best known being Google, with its Google Earth and Google Maps software. This privatization, in addition to acknowledging the loss of an essentially state prerogative since the beginning of the 19th century (Kumar 2010), opens up the field to many other uses of digital maps than those required by the state (see Chapter 6 of this book). Indeed, these devices make digital images of various scales that can be framed in different ways available, thus escaping the constraints and dominant uses of state mapping; they also offer functions that allow maps to be customized according to needs; private software has thus paved the way to participatory mapping, mobilized, for example, in the design of local projects, but also to cartography of contestation or claim aimed at promoting forms of territorial representation alternative to that of the states. One of the best known of these forms of use, and one of the most illustrative of the change in the status of the map in the functioning of state institutions, is Indigenous cartography (see Chapter 7 of this book), which is widely used today to obtain court recognition of land and territorial rights.

More generally, this privatization and diversification of the uses of the map also has the effect of loosening the privileged relationship between the state, the nation and the territory; indeed, it allows collectives other than the nation itself to emerge through or with maps, and even new forms of citizenship. Hence, in line with the hypothesis of this text, which linked the emergence of cartography and the affirmation of the modern state, the present situation allows us to ask whether new forms of social and political order, new forms of authority and territoriality, complementary or alternative to the model of the territorial state, are taking shape (Pickles 2004; Brenner and Elden 2009; Branch 2014; Bailly 2020), by granting contemporary maps with a similar instituting role. But as exciting as this question is, it is beyond the scope of this text, even if it only extends the hypothesis that political forms result from the formation of collective identities in line with frames of reference, especially discursive and visual ones.

5.8. References

Akerman, J. R. (1984). Cartography and the emergence of territorial states in Western Europe. In *Proceedings of the Tenth Annual Meeting of the Western Society for French History*, Sweets, J.F. (ed.). University of Kansas Press, Lawrence.

Agnew, J. (1998). *Geopolitics: Re-Visioning World Politics*. Routledge, New York.

Agnew, J. (2009). *Globalization and Sovereignty*. Rowman and Littlefield, Lanham.

Alliès, P. (1980). *L'invention du territoire*. Presses Universitaires de Grenoble, Grenoble.

Alpers, S. (1983). *The Art of Describing: Dutch Art in the Seventeenth Century*. University of Chicago Press, Chicago.

Anderson, B. (1983). *Imagined Communities, Reflections on the Origin and Spread of Nationalism*. Verso, London.

Arendt, H. (1958). *The Human Condition*. University of Chicago Press, Chicago.

Avila, I. (2015). L'ère des cartes : cartographie, impérialisme et nationalisme en Grande-Bretagne, 1884–1914. In *Les âges de Britannia : Repenser l'histoire des mondes britanniques*, Dunyach, J.-F. (ed.). Presses Universitaires de Rennes, Rennes.

Bagrow, L. and Castner, H.W. (1975). *A History of Russian Cartography up to 1800*. Walker Press, Ontario.

Barber, P. (1992). England II: Monarchs, ministers, and maps, 1550–1625. In *Monarchs, Ministers, and Maps: The Emergence of Cartography as a Tool of Government in Early Modern Europe*, Buisseret, D. (ed.). University of Chicago Press, Chicago.

Biggs, M. (1999). Putting the state on the map: Cartography, territory, and European state formation. *Society for Comparative Study of Society and History*, 41(2), 374–399.

Blais, H. (2014). *Mirages de la carte : l'invention de l'Algérie colonial*. Fayard, Paris.

Blomley, N.K. (2003). Law, property, and the geography of violence: The frontier, the survey and the grid. *Annals of the Association of American Geographers*, 93(1), 121–141.

Boulineau, E. (2008). Fronts et frontières dans les Balkans : les géographes et les enjeux frontaliers sur le Danube en 1919–1920. *Balkanologie*, X(1–2).

Branch, J. (2014). *The Cartographic State: Maps, Territory, and the Origins of Sovereignty*. Cambridge University Press, Cambridge.

Brenner, N. and Elden, S. (2009). Henri Lefebvre on state, space, and territory. *International Political Sociology*, 3(4), 353–377.

Buisseret, D. (1964). Les ingénieurs du roi au temps de Henri IV. *Bulletin de la Section de Géographie*, 77, 13–84.

Buisseret, D. (ed.) (1992). *Monarchs, Ministers, and Maps: The Emergence of Cartography as a Tool of Government in Early Modern Europe*. University of Chicago Press, Chicago.

Buisseret, D. (ed.) (1996). *Rural Images, Estate Maps in the Old and New Worlds*. University of Chicago Press, Chicago.

Buisseret, D. (2007). Spanish peninsular cartography, 1500–1700. In *The History of Cartography*, Harley, J.B. and Woodward, D. (eds). The University of Chicago Press, Chicago.

Carter, P. (1988). *The Road to Botany Bay: An Exploration of Landscape and History*. Alfred A. Knopf, New York.

Casti, E. (2007). State, cartography, and territory in Renaissance Veneto and Lombardy. In *The History of Cartography*, Harley, J.B. and Woodward, D. (eds). University of Chicago Press, Chicago.

Castoriadis, C. (1987). *The Imaginary Institution of Society*. Blackwell, London [Original published in French 1975].

Cohen, S.P. (1971). *The Indian Army: Its Contribution to the Development of a Nation*. UCP, Berkeley.

Conley, T. (1996). *The Self-Made Map: Cartographic Writing in Early Modern France*. University of Minnesota Press, Minneapolis.

Cubitt, G. (1998). *Imagining Nations*. Manchester University Press, Manchester.

Debarbieux, B. (2017). Frontières interétatiques et production des identités collectives. In *Frontières en tous genres*, Staszak, J.F. (ed.). Presses Universitaires de Rennes, Rennes.

Debarbieux, B. (2019). *The Social Imaginary of Space*. Edwin Elgar, Cheltenham [Original published in French 2015].

Debarbieux, B. and Rudaz, G. (2015). *The Mountain. A Political History from the Enlightenment to the Present*. University of Chicago Press, Chicago [Original published in French 2010].

Desbois, H. (2015). *Les mesures du territoire : aspects techniques, politiques et culturels des mutations de la carte topographique*. Presses de l'ENSSIB, Villeurbanne.

Desrosières, A. (2000). L'histoire de la statistique comme genre : style d'écriture et usages sociaux. *Genèses*, 39(2), 121–137.

Didier, E. (2007). Quelles cartes pour le New Deal ? De la différence entre gouverner et discipliner. *Genèses*, 3(68), 48–74.

Dym, J. and Offen, K. (eds) (2011). *Mapping Latin America. A Cartographic Reader*. University of Chicago Press, Chicago.

Edgerton, S.Y. (1975). *The Renaissance Rediscovery of Linear Perspective*. Basic Books, New York.

Edney, M.H. (1997). *Mapping an Empire: The Geographical Construction of British India 1765–1843*. University of Chicago Press, Chicago.

Elden, S. (2013). *The Birth of Territory*. University of Chicago Press, Chicago.

Farinelli, F. (2009). *De la raison cartographique*. CTHS, Paris.

Foucault, M. (2007). *Security, Territory, Population: Lectures at the Collège de France, 1977–1978*. Palgrave Macmillan, London [Original published in French 2004].

Franke, M. (2005). Guerres sans fin : le cas du Nagaland. *Diogène*, 4(212), 85–106.

Gugerli, D. and Speich, D. (2002). *Topografien der Nation: Politik, kartografische Ordnung und Landschaft im 19. Jahrhundert*. Chronos, Zurich.

Hale, J. (2007). Warfare and cartography, ca. 1450 to ca. 1640. In *The History of Cartography*, Harley, J.B. and Woodward, D. (eds). University of Chicago Press, Chicago.

Harvey, P.D.A. (1980). *The History of Topographical Maps: Symbols, Pictures, and Surveys*. Thames and Hudson, London.

Hannah, M.G. (2000). *Governmentality and the Mastery of Territory in Nineteenth-Century America*. Cambridge University Press, Cambridge.

Hansen, J.D. (2015). *Mapping the Germans: Statistical Science, Cartography, and the Visualization of the German Nation*. Oxford University Press, Oxford.

Helgerson, R. (1988). The land speaks: Cartography, chorography, and subversion in Renaissance England. In *Representing the English Renaissance*, Greenblatt, S. (ed.). University of California Press, Berkeley.

Helgerson, R. (1992). *Forms of Nationhood: The Elizabethan Writing of England*. University of Chicago Press, Chicago

Herb, G.H. and Kaplan, D.H. (eds) (1999). *Nested Identities: Nationalism, Territory and Scale*. Rowman and Littlefield, Lanham.

Herz, J.H. (1957). The rise and demise of the territorial state. *World Politics*, 9(4), 473–493.

Hobbes, T. (1651). Leviathan: Or the Matter, Forme, and Power of a Common Wealth Ecclesiasticall and Civil [Online]. Available at: https://www.gutenberg.org/files/3207/3207-h/3207-h.htm [Accessed April 3 2014].

Hopkins, B.D. (2007). The bounds of identity: The Goldsmid mission and the delineation of the Perso–Afghan border in the nineteenth century. *Journal of Global History*, 2(2), 233–254.

Jacob, C. (2006). *The Sovereign Map: Theoretical Approaches in Cartography throughout History*. University of Chicago Press, Chicago [Original published in French 1992].

Jay, M. (1988). Scopic regimes of modernity. In *Vision and Visuality*, Foster, H. (ed.). Bay Press, Seattle.

Johnson, R.G. (1999). *Negotiating the Dayton Peace Accords through Digital Maps*. United States Institute of Peace, Virtual Diplomacy.

Kagan, R.L. and Schmidt, B. (2007). Maps and the early modern state: Official cartography. In *The History of Cartography*, Harley, J.B. and Woodward, D. (eds). The University of Chicago Press, Chicago.

Kain, J.P. and Baigent, E. (1992). *The Cadastral Map in the Service of the State: A History of Property Mapping*. University of Chicago Press, Chicago.

Klein, B. (2001). *Maps and the Writing of Space in Early Modern England and Ireland*. Pelgrave, New York.

Konvitz, J.W. (1987). *Cartography in France, 1660–1848: Science, Engineering and Statecraft*. University of Chicago Press, Chicago.

Krishna, S. (1994). Cartographic anxiety: Mapping the body politic in India. *Alternatives*, 19(4), 507–521.

Kula, W. (1986). *Measures and Men*. Princeton University Press, Princeton [Original published in Polish 1970].

Kumar, S. (2010). Google Earth and the nation state, sovereignty in the age of new media. *Global Media and Communication*, 6(2), 154–176.

Labbé, M. (2010). Les frontières de la nation allemande dans l'espace de la carte, du tableau statistique et de la narration. In *Les espaces de l'Allemagne au XIXe siècle*, Maurer, C. (ed.). Presses Universitaires de Strasbourg, Strasbourg.

Laboulais-Lesage, I. (ed.) (2004). *Combler les blancs de la carte : modalités et enjeux de la construction des savoirs géographiques (XVII–XXe siècles)*. Presses Universitaires de Strasbourg, Strasbourg.

Lascoumes, P. (2007). Gouverner par les cartes. *Genèses*, 3(68), 2–3.

Lefebvre, H. (1991). *The Production of Space*. Oxford, Blackwell [Original published in French 1974].

Lefebvre, H. (2009). *State, Space, World: Selected Essays*. University of Minnesota Press, Minneapolis.

Lukowski, J. (1999). *The Partitions of Poland: 1772, 1793, 1795*. Longman, New York.

Maumi, C. (2007). *Thomas Jefferson et le projet du Nouveau Monde*. Editions de la Villette, Paris.

More, T. (1516). *Utopia*. Thierry Martens, Louvain.

Mundy, B.E. (1996). *The Mapping of New Spain: Indigenous Cartography and the Maps of the Relaciones Geográficas*. University of Chicago Press, Chicago.

Osiander, A. (2001). Sovereignty, international relations, and the Westphalian myth. *International Organization*, 55(2), 251–287.

Ozouf, M.-V. (1989). *La formation des départements la représentation du territoire français la fin du XVIIIe siècle*. Editions de l'EHESS, Paris.

Palsky, G. (1996). *Des chiffres et des cartes : naissance et développement de la cartographie quantitative française au XIXe siècle*. CTHS, Paris.

Pattison, W. (1957). *American Rectangular Land Survey*. University of Chicago Press, Chicago.

Pelletier, M. (2002). *Les cartes de Cassini : la science au service de l'Etat et des régions*. CTHS, Paris.

Pelletier, M. (2007). National and regional mapping in France to about 1650. In *The History of Cartography*, Harley, J.B. and Woodward D. (eds). The University of Chicago Press, Chicago.

Petto, C.-M. (2007). *When France Was King of Cartography: The Patronage and Production of Maps in Early Modern France*. Lexington Books, Lanham.

Pickles, J. (2004). *A History of Spaces: Cartographic Reason, Mapping and the Geo-Coded World.* Routledge, London.

Pinto, J.A. (1976). Origins and development of the Ichnographic city plan. *Journal of the Society of Architectural Historians*, 35(1), 35–50.

Raffestin, C. (1990). *Pour une géographie du pouvoir*. Litec, Paris.

Raj, K. (1997). La construction de l'empire de la géographie, l'odyssée des arpenteurs de Sa Très Gracieuse Majesté, la reine Victoria, en Asie centrale. *Annales. Histoire, sciences sociales*, 20(5), 1153–1180.

Ricci, A. (2015). Maps, power and national identity. The Leo Belgicus as a symbol of the independence of the United Provinces. *Bollettino della Associazione Italiana di Cartografia*, 154, 102–120.

Rousseau, J.J. (1782). *On the Social Contract* [Online]. Available at: www.constitution.org/jjr/socon.htm [Accessed June 15, 2014].

Sahlins, P. (1989). *Boundaries: The Making of France and Spain in the Pyrenees*. University of California Press, Berkeley.

Schama, S. (1987). *The Embarrassment of Riches: An Interpretation of Dutch Culture in the Golden Age*. Alfred A. Knopf, New York.

Schmitt, C. (2006). *The Nomos of the Earth in the International Law of the Jus Publicum Europaeum*. Telos Press, Candor, New York [Original published in German 1950].

Scott, J.C. (1998). *Seeing Like a State. How Certain Schemes to Improve the Human Condition Have Failed*. Yale University Press, New Haven.

Smith, D.K. (2008). *The Cartographic Imagination in Early Modern England: Re-writing the World in Marlowe, Spenser, Raleigh and Marvell*. Ashgate Publishing Ltd, Farnham.

Taylor, C. (2004). *Modern Social Imaginaries*. Duke University Press, Durham.

Ter Minassian, T. (1997). Les géographes français et la délimitation des frontières balkaniques à la conférence de la paix en 1919. *Revue d'histoire moderne et contemporaine*, 44(2), 252–286.

Thomas, F. (2004). The invention of the highlands in Indo-China. Colonial conquest and knowledge production. *Ethnologie française*, 34(4), 639–649.

Tilly, C.E. (1975). *The Formation of National States in Western Europe*. Princeton University Press, Princeton.

Van der Heijden, H.A. (1990). *Leo Belgicus: An Illustrated and Annotated Carto-bibliography*. Alphen aan den Rijn, Canaletto.

Walzer, M. (1967). On the role of symbolism in political thought. *Political Science Quarterly*, 82(2), 191–204.

Williams, A.J. (2011). Reconceptualising spaces of the air: Performing the multiple spatialities of UK military airspaces. *Transactions of the Institute of British Geographers*, 36(2), 253–267.

Winichakul, T. (1994). *Siam Mapped: A History of the Geo-Body of a Nation*. University of Hawai Press, Honolulu.

Wintle, M. (2016). Emergent nationalism in European maps of the eighteenth century. In *The Roots of Nationalism*, Jensen, L. (ed.). Amsterdam University Press, Amsterdam.

6

Statistical Cartography and International Governance in the Age of Big Data

Hy DAO
University of Geneva, Switzerland

6.1. Introduction

Since its advent in the 19th century, statistical cartography has been internationalized both politically and technically. In a context of accelerating and globalizing socioeconomic and environmental changes, and of access to abundant data, we are witnessing a reconfiguration of cartographic production beyond national scales. In the era of Big Data and Open Data, a variety of public and private, national and international actors contribute to cartographies that are supposed to be better adapted to contemporary global territorial and environmental issues. What are the impacts of the technical innovations of recent decades on the actors of international statistical cartography and on the scope of the data collected, processed and visualized? To what political and informational (mis)alignments do these developments lead? This chapter will focus on cartographic production in the context of the sustainable development goals (SDGs).

For a color version of all the figures in this chapter, see www.iste.co.uk/debarbieux/mapping.zip.

The Politics of Mapping,
coordinated by Bernard DEBARBIEUX and Irène HIRT.

6.2. Birth and internationalization of statistical cartography

Statistical or quantitative cartography began to appear in the 19th century, in connection with the development of institutional statistics (Palsky 1996). However, for many years it remained the exclusive domain of the states that possessed the necessary data required for cartographic production. National statistical institutions truly began to develop in the 1830s (Desrosières 2000, p. 123). The first International Statistical Congress was held in 1853 and the International Statistical Institute was created in 1885. However, before the First World War, "international statistics were either the work of individual researchers [...] or of compilers holding high positions in national and international statistical institutions [...], or of national official bodies" (Etemad 1994). International statistics began as an activity of transnational cooperation of a scientific nature before, in 1919, it was structured in an international organization in the Economic and Financial Organization (EFO) of the League of Nations (LON): "With the EFO, international statistics, based on comparison between states and supported by governmental agreement, become an original government tool: it directly unites the ethics of objectives (peace) with quantitative knowledge without pluralist mediation" (Cussó 2012, p. 130). We are thus witnessing the creation of a technical instrument overlooking the states, giving itself the ambition of working beyond the specific interests of nations: "the LON quantifies with governments, but without states" (Cussó 2012, p. 119).

Several international organizations began publishing their data during the interwar period: statistics from the International Labour Organization in 1919, the first edition of the International Statistical Yearbook by the League of Nations in 1927, the first world agricultural census in 1930 by the International Institute of Agriculture (IIA), the future Food and Agriculture Organization (FAO) of the United Nations.

After the Second World War, the movement grew with the regular publication of statistics by international institutions such as the FAO (1945), the United Nations Statistical Office (1946) or the World Bank (1947), in a context of interest in questions of economic development, resource use and food. Within the system of international organizations, the central role of the United Nations is apparent from the first agreements with the specialized agencies, such as that of 1946 with the FAO[1], in which the right of the United Nations to produce data in the field of the FAO is specified:

1 Available at: https://www.unsystem.org/content/agreements-between-specialized-agencies-and-united-nations.

> 2. The Food and Agriculture Organization of the United Nations recognizes the United Nations as the central agency for the collection, analysis, publication, standardization, and improvement of statistics serving the general purposes of international organizations.
>
> 3. The United Nations recognizes the Food and Agriculture Organization of the United Nations as the appropriate agency for the collection, analysis, publication, standardization and improvement of statistics within its special sphere, without prejudice to the right of the United Nations to concern itself with such statistics so far as they may be essential for its own purposes or for the improvement of statistics throughout the world. (Article XII of the 1946 agreement between the United Nations and the FAO).

The right of the FAO to deal with statistics outside its sphere of competence is not specified in this document. Although some of the agreements with other agencies will later contain the following statement in their favor: "without prejudice to the right of the [*the agency*] to concern itself with any statistics so far as they may be essential for its own purposes", they nonetheless recognize the prominent place that the central bodies of the United Nations (notably the Statistical Commission and the Statistics Division) will have in the international statistical system.

6.3. International statistical mapping

6.3.1. *Statistics and mapping by international organizations*

From the second half of the 20th century onwards, international organizations (the United Nations with its agencies, funds and programs) were therefore the main actors in the production of international statistics, within the framework of their missions and in collaboration with states.

The FAO has been a pioneer not only in statistics but also in cartography. As a United Nations agency specializing in food and agriculture, the FAO has conducted periodic inventories of the world's forest resources since 1948[2] at intervals of about ten years on the basis of data provided by countries, but also remote sensing. In the first report, from 1948, one can note the absence of cartographic presentation of the results but, on the other hand, the presence of an important chapter devoted to the possibilities of aerial photography "for speeding up and putting on a uniform basis the knowledge of the forests of the world that is now much needed" (FAO 1948).

2 Available at: http://www.fao.org/forest-resources-assessment/past-assessments/.

The use of aerial images is promoted for technical purposes, but also to obtain information that is independent of that provided by the states. From the 1970s onwards, and particularly for the FAO's 1980 assessment of tropical forest resources (Lanly 1982), satellite images – mainly from the US Landsat program – were systematically used to complement national statistics. The use of remote sensing marks a departure from the traditional state limitations of statistical mapping. Not only do these data sources allow independent evaluations of national statistics (as long as access to the data is allowed by the state(s) controlling the Earth observation programs concerned), but the spatial units of observation become flexible, and can be broken down into specific transnational perimeters (e.g. watersheds, protected areas).

The FAO develops cartographic expertise internally, but also relies on collaborations. Starting in 1961, following a recommendation by the International Union of Soil Sciences, the FAO (together with the United Nations Educational, Scientific and Cultural Organization – UNESCO) was the originator of one of the first global environmental maps, the *Soil Map of the World*[3], the result of a two-decade collaboration with many soil scientists around the world (see Figure 6.1).

With the launch of a statistical database on agriculture in 1963, the FAO produced innovation in the field of information systems. This database was followed by the launch of AGROSTAT, which has now become FAOSTAT[4], in 1986, a thematic reference source for international statistical mapping.

In the environmental field, the United Nations Environment Programme (UNEP) has carried out numerous data and mapping activities since its creation in 1972, in accordance with its mandate "to keep under review the world environmental situation in order to ensure that emerging environmental problems of wide international significance receive appropriate and adequate consideration by Governments" (United Nations Resolution 2997, 15 December 1972). Occasionally, cartographic activities have been delegated to specialized centers or experts, as in the case of the GLASOD global land degradation inventory of the early 1990s, the result of the collaboration of more than 250 national experts coordinated by a scientific foundation, ISRIC (International Soil Reference and Information Centre). However, UNEP has also developed its mapping skills internally. In 1985, the Global Resource Information Database (GRID) network was established, which, in

3 Available at: http://www.fao.org/soils-portal/data-hub/soil-maps-and-databases/faounesco-soil-map-of-the-world/en/.

4 Available at: http://www.fao.org/faostat/.

the 2000s, developed the Environmental Data Explorer (EDE)[5], the reference portal for statistical data on the environment in support of the Global Environment Outlook series, UNEP's flagship reports on the state of the world's environment.

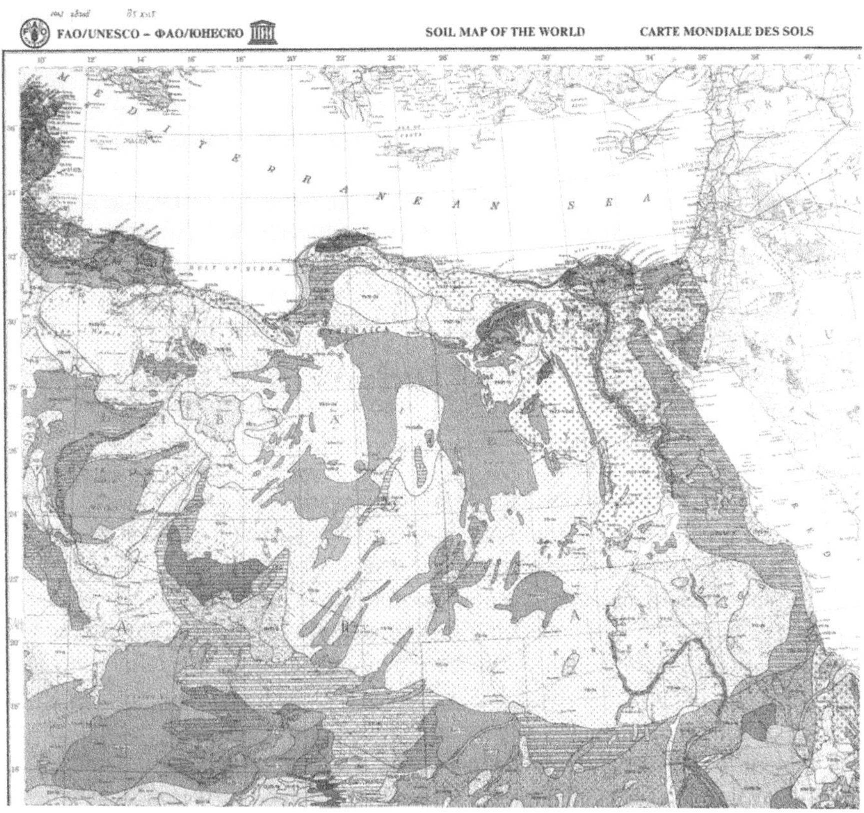

Figure 6.1. *FAO/UNESCO Soil Map of the World (edition 1/1973, Africa, from sheet VI-2) (source: Food and Agriculture Organization of the United Nations. Reproduced with permission)*

In the area of development, the World Bank's objectives are to support countries and the UN system in providing statistical data in order to monitor poverty reduction strategies. The compiled and harmonized data are made available openly and free of

5 Available at: http://ede.unepgrid.ch/.

charge through an interactive[6] data portal and a mapping application[7] that also displays information on World Bank–supported projects.

In addition to the abundance of data and maps produced by each organization, coordination tools have been developed to strengthen the coherence of geostatistical information in the international system.

6.3.2. *Coordination of statistical and geographical information*

With regard to statistical data, the United Nations Statistics Division (UNSD) within the Department of Economic and Social Affairs is in charge of the global statistical system. It works with international organizations, supports countries in strengthening their national systems, compiles and disseminates reference data, and develops statistical and geographic standards (including country codes and names[8]). The UNSD maintains the UN data portal[9], launched in 2005 as part of the *Statistics as a Public Good* open data project. This brings together data from many institutions and allows them to be downloaded, but does not offer cartography.

For geographic information, the Global Geospatial Information Management (UN-GGIM, established in 2011) provides coordination within the UN system. The United Nations Geospatial Information Section (formerly the United Nations Cartographic Section) is the unit responsible for providing cartographic products for the UN Secretariat. In particular, this section provides the official UN Clear Map dataset in order to facilitate "the issuance of any map at any duty station, including dissemination via public electronic networks such as Internet"[10]and "to ensure that maps meet publication standards and that they are not in contravention of existing United Nations policies"[10]. One of the purposes of publishing these reference maps is to avoid complaints from countries affected by disputed territorial boundaries or names. For example, national boundaries on seas (including inland seas) are not shown on UN Clear Maps (see Figure 6.2).

6 Available at: https://data.worldbank.org/.

7 Available at: https://maps.worldbank.org/.

8 Through the United Nations Group of Experts on Geographical Names, available at: https://unstats.un.org/unsd/ungegn/.

9 Available at: https://data.un.org/.

10 Available at: https://geoservices.un.org/.

Figure 6.2. *Absence of national boundaries on maritime territories (United Nations) (source: https://geoapps.dfs.un.org/Html5Viewer/Index.html?viewer=Clearmap# [Accessed 16 November 2020])*

NOTE.– The preparation and use of these seemingly simple background maps receives considerable attention both from states and the United Nations, particularly in territories that are subject to political tension or conflict.

For Earth observation data, the lead entity is the Group on Earth Observations (GEO), a structured partnership of more than 100 member states and 100 participating organizations (international and non-governmental). It is coordinating the development of a Global Earth Observation System of Systems (GEOSS): "to better integrate observing systems and share data by connecting existing infrastructures using common standards"[11].

6.3.3. *Is the aggregation of national data outdated?*

For several decades, international organizations have been developing their capacities to produce maps and, beyond that, to generate and manage the data required for these maps. These capacities have been put in place by each of these organizations, initially to meet their own needs, with sometimes a proliferation of cartographic or geomatic projects within the organizations themselves, depending on specific initiatives or funding. There is a great deal of coordination within the United Nations in terms of statistics and background maps (since 1945), as well as

11 Available at: https://www.earthobservations.org/.

geospatial information (since the 2000s). On the other hand, apart from the cartographic templates of the United Nations Secretariat or those established within each organization (such as the Office of the United Nations High Commissioner for Refugees), the production of data and maps by international organizations is very diverse, covering a wide range of United Nations fields of action (agriculture, environment, development, health, humanitarian, risks) and responding to various functions (providing harmonized and interoperable statistical data, operational information, monitoring and analysis, illustration of reports, evaluation of programs or policies).

International statistics are still mostly constructed through aggregations of data provided by countries to intergovernmental organizations. The basic observation units therefore remain within the borders of the states, even though there are many examples of sub-national (notably obtained by remote sensing or by spatial disaggregation) or transnational (collected or put together according to perimeters that are independent of state boundaries, for example, watersheds) global data. These observations of the territory through infra- or transnational spatial units, even if they are intended to remain within the framework of agreements between governments, do not always escape the constraints of the structuring and dissemination of data by the states[12]. Changes in international statistical and cartographic production are, however, taking place in the contemporary context of the widespread digitization of the quantitative explosion of information.

6.4. Cartography in the digital age

6.4.1. *Computerization and dissemination of cartographic resources*

6.4.1.1. *The computerized turning point of cartography*

Simultaneous with the development of international organizations, the post-war period saw the beginnings of computer technology and its application to new approaches to cartographic production, "automatic" computer cartography (Tobler 1959), and geographic information systems (GIS, a term proposed in 1962 by Roger Tomlison in order to designate a cadastral inventory system in Canada). The year 1972 was a turning point for international cartography with the US space agency, the National Aeronautics and Space Administration (NASA), commissioning the first civilian Earth observation satellite, Landsat, which revolutionized data collection

12 See, for example, the unfinished implementation of the *Système d'observation et d'information des Alpes* (Alpine Observation and Information System) – SOIA, notably for political and technical issues of data access (Pieroni and Debarbieux 2014).

with its spatial resolution and spectral richness, which was far superior to those of the meteorological satellites that had been available since 1960. Landsat ushered in an era of continuous observation of our planet and an unprecedented accumulation of temporal data series. For 15 years, the United States maintained its hegemony in Earth observation, until the appearance of Soviet (Resurs), French (SPOT) and Indian (IRS) satellites, in the 1980s.

It was also the 1970s when the digitization of registers, cadastres and statistical inventories began to become widespread in administrations. The diffusion of cartographic tools on microcomputers began in the 1980s, which saw a blossoming of software that was dedicated to statistical cartography (particularly in France and the United States). The first web mapping technologies appeared in the 1990s. Despite these increased possibilities for cartographic production, particularly for non-governmental actors such as researchers or private enterprises, data are still relatively inaccessible and require considerable preparation before they can be used for statistical mapping: censuses and administrative registers are not systematically geocoded, data are rarely provided in a digital format and background maps are not always available or matched with data tables.

6.4.1.2. *Beginning to open*

Nevertheless, governmental decisions toward greater openness of geographic data are beginning to be taken. In the 1980s, the civilian use of the global positioning system (GPS), which was initiated in 1978, was authorized. In May 2000, the US government put an end to the intentional degradation of the accuracy of GPS data (selective availability). In the meantime, other positioning systems have been developed by Russia (Glonass), Europe (Galileo) and China (Beidou) to avoid dependence on the United States.

In the 2000s, after some 20 years of marketing strategies via private companies, Landsat images were made available free of charge, with the stated aim of maintaining American leadership in remote sensing, which can be understood by the symbolic value gained by the widespread use of a product but also by the price competition against other possible initiatives by countries with fewer financial resources. At the same time, the mapping tools *Google Maps* (2004) and *Google Earth* (2005) were launched, which are today the two most widely used mapping applications in the world by far. These two steps mark the diffusion of geographic information outside the state services (NASA and the United States Geological Survey, USGS in the United States for Landsat images) as well as a decoupling between national sovereignty territories and observation spaces (which can go as far as including the entire globe).

6.4.1.3. *Data deluge and web access since the 1990s*

The hegemony of the United States over the last few decades, due to its capacity for technological innovation but also to its policy of data dissemination, is being undermined by the multiplication of sources and the increasing accessibility of data and computer resources worldwide. Indeed, since the advent of the Internet, and more particularly from 1993 onwards, the year in which the European Organization for Nuclear Research (CERN) placed web software in the public domain, cartography has been turned upside down by the explosion of data and its availability on the Web.

According to the World Bank[13], in 2017, almost half of the world's population used the Internet, although there are still significant disparities between continents and countries. In 2012, IBM stated that "90% of the data in the world today has been created in the last two years"[14]. Forecasts by Reinsel *et al.* (2018) indicate that by 2025 the datasphere is expected to increase more than fivefold, from 33 to 175 zettabytes[15]. Another remarkable development concerns the share of real-time data (from machines, sensors, connected objects), which could double from 2017 to 2025 to represent 30% of the "global datasphere" (Reinsel *et al.* 2018). These trends are leading to the establishment of a new global information space, which is based on Internet technologies in which international cartography is now unfolding.

6.4.2. *A global information space based on the Web and Big Data*

6.4.2.1. *A global web-based data infrastructure*

Although the use of web data for cartographic purposes and statistical or spatial analyses is still in its infancy, geo-referencing is already largely a determining organizing principle of a new global information space, the Geoweb. It is "a spatial organization of information on the Internet through direct or indirect georeferencing on the Earth's surface" (Joliveau 2010). The Geoweb is built on a multiform material infrastructure of data acquisition, storage, analysis, dissemination and consultation that is composed of sensors, computer servers, networks, personal computers or even telephones (Joliveau 2010).

---

13 Available at: https://data.worldbank.org/indicator/IT.NET.USER.ZS.

14 Available at: https://www.facebook.com/IBM/posts/90-of-the-data-in-the-world-today-has-been-created-in-the-last-two-years/293229680748471/.

15 33 x 1021 bytes, that is, the storage capacity of 4,000 billion DVDs or the equivalent of 500 DVDs for each inhabitant of the Earth.

As a result, the number of data sources available on the Web for cartography is increasing (see Table 6.1).

Conventional data	Organized geographic information	Statistics, earth observation, ground measurement stations
Non-conventional data	Voluntary geographic information (Goodchild 2007)	Crowdsourcing, citizen science, participatory GIS = active contribution
	Unintentional geographic information (Fischer 2012)	Social networks, cell phone traces, Internet of Things = passive contribution

Table 6.1. *Conventional and non-conventional data for cartography*

The Geoweb is therefore developing thanks to the technical innovations of the Web and measurement systems. The Web makes it possible to set up ergonomic interfaces for interactive access to maps and data (Webmapping, virtual globes[16] and story maps[17]). However, between "managerial optimization and democratic requirements" (Gautreau and Noucher 2013), the Geoweb also extends through a political will of governments at all scales to make the use of data more efficient and transparent in order to respond to contemporary environmental and development challenges characterized by their thematic multidimensionality, their global extent, their rapid evolution and their interdependence.

6.4.2.2. *New transnational data worlds*

In the Geoweb, and following Buckland who wrote, before the advent of the Web, that "everything is, or might well be, information" (Buckland 1991, p. 356), one can affirm that everything is or could be geo-information and therefore a map. The accumulation of data precedes intention, a situation that is radically different from traditional approaches that proceeded through statistical collection programs established according to previously identified needs. Because of these continuously acquired data, such as satellite images, cell phone traces, exchanges on social networks, measurements by connected objects (watches, cameras, household

16 See, for example, the Earth application https://earth.nullschool.net/fr/, which provides a systemic view of the global environment by combining numerous global datasets from different sources. Political territories are absent from this visualization.

17 A web application combining maps and multimedia elements in a narrative thread with which the user can interact (navigate within the text, access maps, data, graphics, movies, etc). Political institutions are increasingly interested in these cartographic storytelling tools to stage their information in an attractive way for communication purposes.

appliances, televisions, cars), a new potentially visualizable invisible – which escapes state statistics and its thematic, spatial and temporal categories – can be revealed.

Remote sensing data allow transnational world views, with high spatial (down to less than 1 m) and temporal resolution. Thus, the global image generated from MERIS sensor data (see Figure 6.3) is a composite of images taken at different dates and produced by geometric and radiometric correction calculations. Presenting itself as a natural "true color" image, it nevertheless overwrites physical temporality (the differences between days, seasons and years) and does not show boundaries of political territories in order to highlight an environmental issue. In this, it is no less political than any data created with a predefined intention.

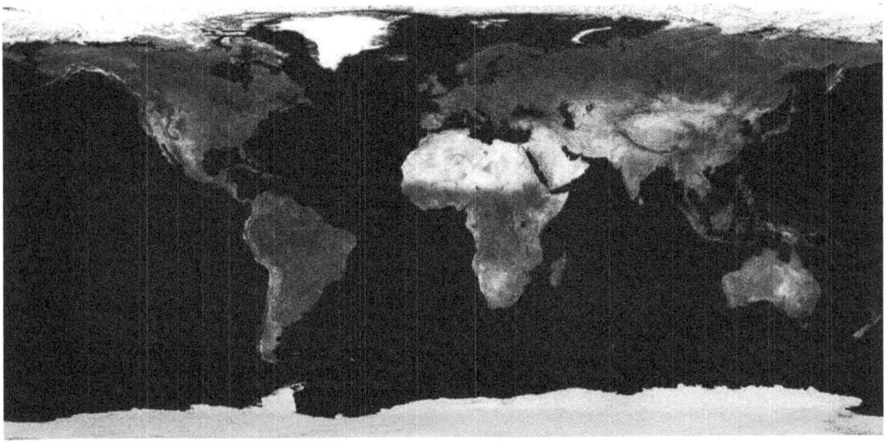

Figure 6.3. *Global mosaic of images from the MERIS sensor (ESA) "This mosaic of global land cover images taken by the MERIS instrument onboard ESA's environmental satellite shows the entire planet 'in its true colors'. To produce the mosaic, a total of 1,561 satellite orbit passes were used over the period May–November 2004" (ESA, CC BY-SA 3.0 IGO, source: https://www.esa.int/ESA_Multimedia/Images/2006/06/Mosaic_of_global_land_cover_images_taken_by_MERIS#.XkqSPQyrU-Y.link [accessed 16 November 2020])*

As an example of unconventional data, the traces collected by the Strava sports application[18], translated into heat maps, provide continuously updated information on certain mobilities, particularly cycling (Lee and Sener 2020), observed

18 Available at: https://www.strava.com/heatmap.

independently of political territories, and are complementary to statistics from censuses or surveys.

6.4.2.3. *Spillover from Big Data*

With the multiplication of geodata sources combined with an exponential growth of computer systems capabilities, the Geoweb has entered the era of Big Data, a term that emerged in the 1990s in the context of database management. Big Data, characterized by the 5 Vs (Volume, Velocity, Variety, Value, Veracity)[19], refers to data sets that are too large and complex to be analyzed with traditional statistical methods. In the age of the Internet, Big Data includes a variety of sources (see Table 6.2).

Traditional organizational systems: data from processing	Commercial or banking transactions, medical records, etc.
Social networks: man-made information	Social networks such as Facebook, Twitter, instant messaging systems, image/video exchange systems, Internet search engines, e-mails, etc.
Internet of Things: machine-generated data	Ground or airborne/satellite sensors, mobile phones, GPS, computer logbooks, etc.

Table 6.2. *Types of Big Data (after: UNECE High-Level Group for the Modernisation of Official Statistics 2013)*

Big Data are produced by both public and private activities for a wide variety of purposes (commercial, surveillance, leisure, scientific, etc.). The volume of data generated is such that it requires storage and processing capacities that public organizations are often not able to provide themselves or that they prefer to outsource to private companies, especially technology companies and web giants such as GAFAM (Google, Apple, Facebook, Amazon and Microsoft). These companies are increasingly developing mapping services, both for their own needs and public entities. For example, Facebook's *Data for Good*[20] initiative produces global population data at 30 m spatial resolution, based on population statistics combined with satellite imagery and processed by artificial intelligence (machine learning) (see Figure 6.4).

---

19 An extension of a first definition that defined 3 Vs (Volume, Velocity, Variety) by Laney (2001).

20 Available at: https://dataforgood.fb.com/tools/population-density-maps/.

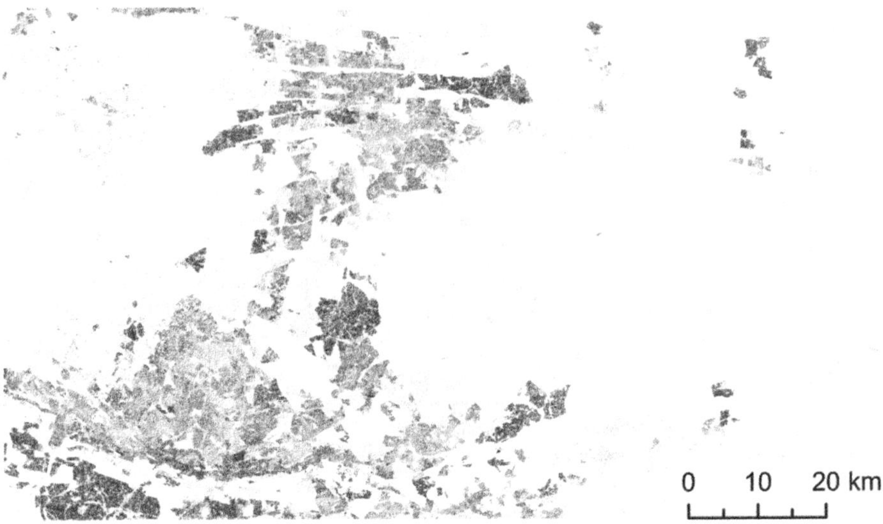

Figure 6.4. *World population map at 30 m spatial resolution, zooming in on Johannesburg and Pretoria, South Africa (Facebook Data for Good) (source: Facebook Connectivity Lab and Center for International Earth Science Information Network* et al. *2016 HRSL source imagery © 2016 DigitalGlobe*

While these maps initially serve the company's own needs (to evaluate the accessibility and coverage of the Internet by different populations), they are now made available to international organizations with the stated objective of helping to solve some of the world's greatest problems. This type of product from private actors disrupts institutional data: on the one hand, it refines existing statistics, but on the other hand, the information generated has no official validation; it is built on methodologies that are not necessarily transparent and it has no guarantee of accessibility and comparability over the long term.

Another illustration of the insertion of GAFAMs into the cartographic production of international organizations is the GEO's approach to partnerships with Amazon and Google. These partnerships[21] allow international organizations to access the processing power of the companies involved, but are problematic in terms of dependence on external and private systems and data ownership.

21 GEO–Amazon Earth Observation Cloud Credits Programme (available at: http://www.earthobservations.org/article.php?id=369), GEO–Google Earth Engine Programme (available at: http://www.earthobservations.org/geo_blog_obs.php?id=406).

6.4.2.4. *Opening, standardization and sharing of data*

Access to Geoweb data also refers to the principles of openness, standardization and mutualization. The idea of open data has been promoted by scientists since the 1950s, but it was in the 1990s, alongside the emergence of the Web and Big Data, that the concept of Open Data emerged and began to spread more widely in society, advocating the principle that "open data and content can be freely used, modified, and shared by anyone for any purpose" (Open Knowledge Foundation)[22]. Since the 2000s, governments have embraced the concept in response to a societal demand for transparency, but also for the purpose of promoting economic activities around geodata. Following the lead of pioneering countries such as the United States and France, many countries now have Open Data strategies and action plans aimed at ensuring that governments make a concrete commitment to opening up the data they produce, commission or hold. In this way, Open Data marks governments' abandonment of the idea of exclusive use of data for their own purposes.

As for the standardization of geo-information, which is necessary for the interoperability of systems, it goes back to the first national and international geographic data infrastructure initiatives in the 1990s (standards of the Federal Geographic Data Committee in the United States, as early as 1992, and of the ISO/TC 211 technical committee "Geographic Information/Geomatics" as early as 1994). Currently, general specifications in the field of geomatics information are mainly published by the Open Geospatial Consortium (OGC)[23], an international consortium of 500 government agencies, companies and universities. Despite the differences, and even competition, between standards (international and national), they now allow the effective interconnection of databases between organizations.

Openness and standardization are tools to allow the pooling of information within and between administrations for reasons of economy, but also to enhance the value of the information capital represented by data that are largely financed by public funds. This pooling takes the form of complex infrastructures that are made up of geodata and IT tools, but also methodologies, standards and institutions. These spatial data infrastructures have acquired a political existence not only within states but also at the international level, particularly in the UN system, and even at the supranational level, such as the INSPIRE directive (2007) that is implemented in the European Union. Within administrations, these geo-informational infrastructures generally have a cross-disciplinary place in relation to other policy areas, creating spaces for not only technical but also political dialogs concerning information strategies for organizations.

22 Available at: https://opendefinition.org/.

23 Available at: https://www.ogc.org/.

The impact of this new global information space on international statistical mapping will be illustrated in the next section, using the example of the UN SDGs indicators.

6.5. Mapping the SDGs

6.5.1. *A set of non-binding targets for 2030*

Since 2015, the SDGs have been the global development reference framework for most international, but also national and local, policies. The SDGs are a set of 17 goals (broken down into 169 targets) adopted by the Member States of the United Nations after a broad consultation process also involving other actors and gathered in the 2030 Agenda. Unlike the 8 millennium development goals (MDGs) adopted in 2000, which were aimed at developing countries, the SDGs concern all the countries of the world. The SDGs do not have a legally binding status and rely on the voluntary involvement of states, which have, nevertheless, generally made a strong political commitment to adopting the SDGs in the design, conduct and monitoring of their policies.

Although the MDGs were primarily socioeconomic and defined in a top-down manner by a small number of people within the United Nations, the MDGs cover more areas, especially the environment, and are the result of a broader consultation process that also includes businesses, civil society and scientists. The 2030 Agenda and the SDGs aim to cover all areas and countries, and to "leave no one behind" (point 4 of the 2030 Agenda). They are subject to structured monitoring based on statistical indicators.

6.5.2. *Governance of the SDG indicators*

The High-Level Political Forum on Sustainable Development, including Member States, is the political platform for the follow-up of the 2030 Agenda. However, it is the Inter-agency and Expert Group on SDG Indicators (IAEGs–SDGs)[24] that is responsible for developing and implementing the global framework of SDG indicators. This framework was adopted in 2017, and the SDG indicators are regularly revised through processes that include open consultations with civil society, researchers and businesses. The 232 SDG indicators (as of end 2019) are used as the basis for the annual SDG progress report (United Nations 2019).

24 Available at: https://unstats.un.org/sdgs/iaeg-sdgs/.

The compilation and verification of indicators is the responsibility of UN entities and international organizations, the Custodian Agencies[25] (see Figure 6.5), which interact with national agencies and submit, after validation by the relevant countries, data to the central database of SDG indicators maintained by the UNSD[26].

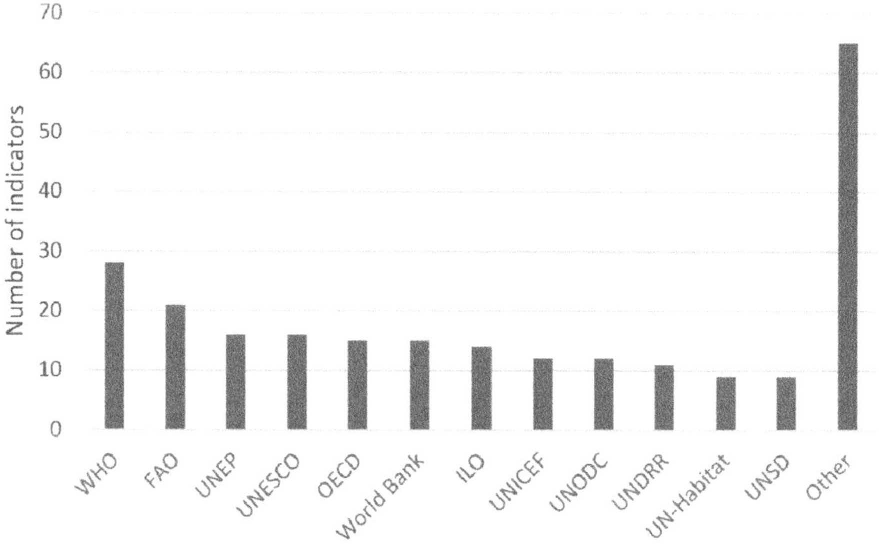

Figure 6.5. *Main Custodian Agencies of the SDG indicators*

6.5.3. *Information coverage of the SDGs*

Indicators are categorized into three levels (tiers) based on clarity of definition, existence of methodology and availability of data, with Tier 1 indicators being those meeting all criteria[27]. Of the 232 SDG indicators, 116 indicators are Tier 1, 92 are

25 See, for example, indicator SDG 6.6.1, for which UNEP is responsible, which has developed a mapping information platform on this SDG, in partnership with the European Commission's Joint Research Centre (JRC) and Google, available at: https://www.sdg661.app/.

26 Global SDG Indicators Database, available at: https://unstats.un.org/sdgs/indicators/database/.

27 Tier 1: Indicator is conceptually clear, has an internationally established methodology and standards are available, and data are regularly produced by countries for at least 50% of countries and the population in every region where the indicator is relevant. Tier 2: Indicator is conceptually clear, has an internationally established methodology and standards are available, but data are not regularly produced by countries. Tier 3: No internationally established methodology or standards are yet available for the indicator, but methodology/standards are being (or will be) developed or tested (more information available at: https://unstats.un.org/sdgs/iaeg-sdgs/tier-classification/).

Tier 2, 20 are Tier 3 and 4 are of multiple tiers (status end 2019). Figure 6.6 shows the uneven distribution of the SDGs.

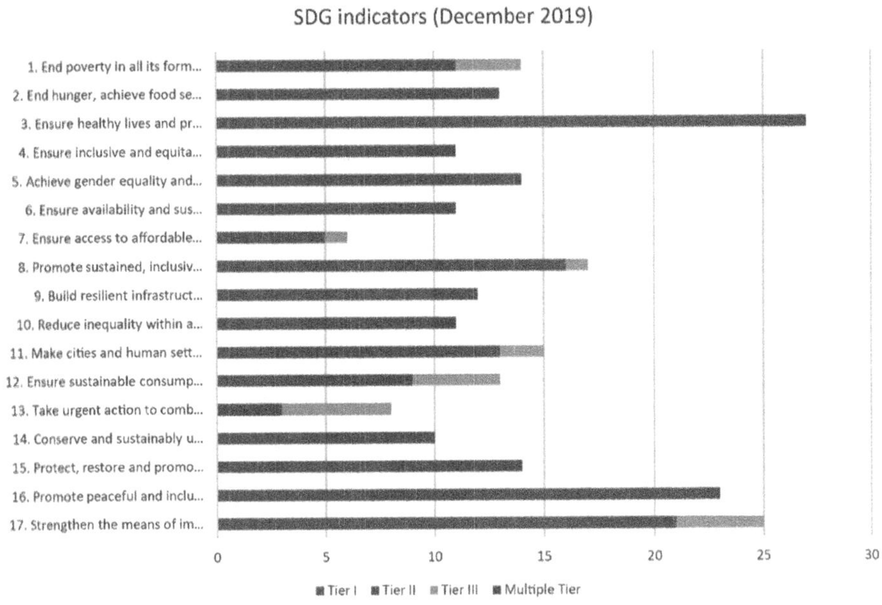

Figure 6.6. *Levels (tiers) of SDG indicators, by SDG*

The number and robustness of the indicators, in conjunction with the availability of data, obviously influence the perception of the issue at stake as well as the assessment of country performance. While the indicators are subject to multiple discussions and revisions at both political and technical levels, the Custodian Agencies, through their technical expertise and authority in their respective fields, have a central role in defining the measurement of the SDGs.

6.5.4. *Cartographic visualizations of the SDGs*

The SDG indicators are made available through the Global SDG Indicators Database (see footnote 26) maintained by UNSD as well as through web[28] services that allow direct use of the data in third-party IT applications. The indicators are mainly provided in table form as a systematic mapping of all indicators is not

28 Available at: https://unstats.un.org/SDGAPI/swagger/.

currently available. The Open SDG Data Hub[29] launched by UNSD provides a number of geo-services, but the general approach seems to be to provide data that is suitable for analysis and mapping by its users rather than the indicator maps themselves. This is symptomatic of a contemporary conception of cartographic production, which is open, shared and even left to actors other than the data provider (whereas many statistical offices around the world have a tradition of statistical cartography, sometimes jealously guarded with arguments of statistical competence or secrecy).

In general, it can be seen that the SDG indicators are relatively poorly represented in a map form. A reference report such as the Global Sustainable Development Report 2019, mandated by the United Nations (Independent Group of Scientists appointed by the Secretary-General 2019), does not contain any maps, while the Sustainable Development Goals Report 2019 (United Nations 2019) presents only a small number of maps, in the form of thumbnails, which are in the minority when compared to other visual representations (graphs, images). Among the examples of mapping identified, especially online, there are recurrent mapping errors, such as disproportions of areas from certain projections (especially Mercator) or the use of inappropriate visual variables[30] (Kraak *et al.* 2018).

The limited use of cartography in the case of the SDGs, even though it allows for synthetic views of global situations and numerous data are available, is undoubtedly explained by multiple factors, notably linked to a statistical culture that is primarily concerned with the elaboration of methodologically validated data, as well as to a more widespread use of visualizations by means of graphs, which are less complex to produce than maps and more legible for policies in terms of progress toward the goals.

The dashboard accompanying the Sustainable Development Report, published by the Sustainable Development Solutions Network and the Bertelsmann Foundation (Sachs *et al.* 2019), is an interesting example of mapping the SDG indicators in terms of goal achievement. The maps available online[31] from this dashboard show scores resulting from a specific calculation methodology, making the information both more abstract, subject to questioning and yet potentially more usable by policy actors.

29 Available at: https://unstats-undesa.opendata.arcgis.com.

30 In particular, absolute quantities (e.g. population totals) are shown as choropleths when they should be shown as proportional symbols.

31 Available at: https://dashboards.sdgindex.org/.

Despite the openness of current data, it is still difficult to find general maps of the SDG indicators that could have been used by actors outside the United Nations, as the Gapminder Foundation did for the MDGs with its innovative visualizations (animated time series, graphs inlaid in presentation videos)[32]. Is it because the indicators and SDG data platforms are still new? Is it because of a lack of awareness of the analytical added value that maps can bring? Is it a lack of interest in map galleries that present data for illustration purposes without interpretation? These questions remain open.

6.5.5. *The SDG indicator process challenged by Big Data*

For the time being, the SDG indicators are mainly quantified by aggregating national statistics. The role of Big Data, in particular in better covering themes and populations that are still insufficiently taken into account, is recognized by the United Nations (UN/IEAG 2014)[33]. Applications of Big Data for mapping on a global scale are emerging (see Figure 6.7)[34].

The potential of Big Data is estimated to be particularly high for the measurement of SDGs 3 (health), 8 (labor) and 11 (cities) (UN Global Pulse 2017; MacFeely 2018). Analytics from Big Data can complement information on realities insufficiently covered by statistics, as shown by the study by Le Goix *et al.* (2019) on housing affordability in Europe, which employs web scraping techniques to acquire current and detailed information on the housing market (SDG indicator 11.1 on access to affordable and quality housing).

The use of Big Data for international statistical mapping is still at an exploratory stage, but it already poses various challenges to institutions in terms of legislation (access and usage rights), privacy (use of personal data), finances (data acquisition costs, especially from private sources), management (resources, data protection), methodology (data quality, statistical methods) and technology (processing

32 Available at: https://www.gapminder.org/.

33 See also the work of the UN Global Working Group on Big Data for Official Statistics (available at: https://unstats.un.org/bigdata/).

34 See also other examples, such as the analysis of tweets on climate change (UN Global Pulse, available at: http://www.unglobalpulse.net/climate/), the analysis of deforestation through massive satellite image processing (Global Forest Watch, available at: https://www.globalforestwatch.org/map) or the detection of fishing activities (Global Fishing Watch, available at: https://globalfishingwatch.org/our-map/).

capacities)[35]. Cartography based on Big Data could, in the future, constitute entirely alternative solutions to existing statistical methodologies. For example, Giuliani *et al.* (2020) have developed a global data and processing workflow for the SDG 15.3.1 indicator "Land degradation", for which little data are currently reported by the United Nations. In cases where there is competition between different quantifications, the approach taken will depend on negotiations that will be as technical as they are political, especially if the figures produced differ from national data.

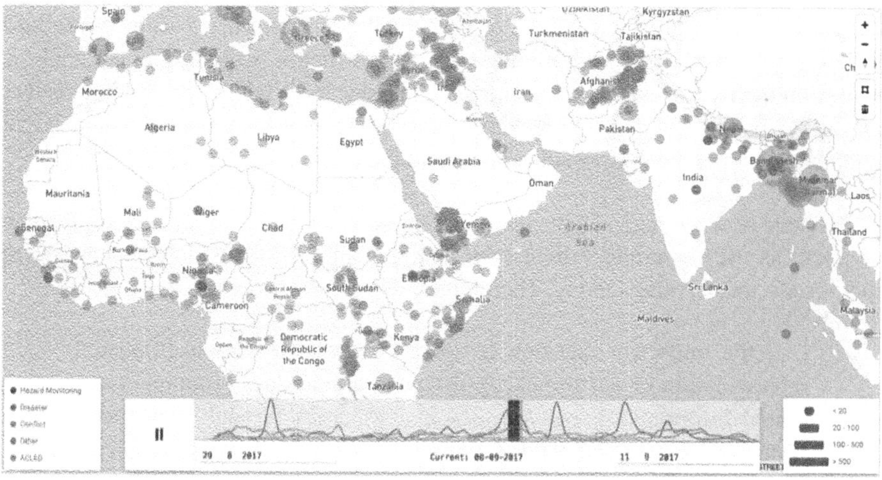

Figure 6.7. *Automatic analysis of media texts on internally displaced people (source: Internal Displacement Monitoring Centre (IDMC), available at: https://www.internal-displacement.org/monitoring-tools/monitoring-platform)*

6.6. (Re)configurations of international cartography

In this global context of quantifying policy monitoring and the new possibilities of the Geoweb and Big Data, international cartography is being impacted functionally, informationally and politically.

6.6.1. *From the visual paradigm to the datamatic paradigm*

With the advent of digital technology, the map paradigm is at least as much "datamatic" – based on the logic of data processing – as it is visual. The map,

35 Commission économique pour l'Europe (2013). Utilisation des "données massives" dans les statistiques officielles. *Conference of European Statisticians*, March 2013.

generated from an information system, is a "visualizable file (which) preserves the system of relations of the map without the image" (Raffestin *et al.* 1995), visualization being only one mode of using the information, among others. The map's function of "presenting the known" (MacEachren and Kraak 1997) is weakened in favor of exploration. In fact, there is a certain esthetic and graphical poverty of internet maps, when it is not a question of semiological errors (Kraak *et al.* 2018). Exploration through the map refers first to its use as an interface for accessing information, rather than as an accomplished visual product: digital maps "literally act like windows, not pictures" (Jennings 2011, p. 213). This map-interface can also be backed up by data mining tools from which it is used to render analysis results on the fly, for example, business intelligence tools with their dashboards that are much prized by decision-makers. This is the non-predefined, dynamic and interactive map of the geo-visualization paradigm (Crampton 2001, p. 238).

Esthetic or semiological considerations are set aside; it is the capacity of the map to allow its user to produce their own analyses and representations, according to their needs, that is valued today. Therefore:

> [...] the important question is not is not what a map is (a spatial representation or performance), nor what a map does (communicates spatial information), but *how the map emerges* through contingent, relational, context-embedded practices to solve relation problems (their ability to make a difference to the world). (Kitchin and Dodge 2007, p. 342)

It can be observed that the maps of international organizations, which are present in large numbers on the Internet or in publications, are more to be constructed than built, often more illustrative of data than conveying messages.

6.6.2. *Informational alignments*

The multiplication of international cartographic productions, thanks to the access to a growing number of technologies and data sources, questions the role of the map as "a common reference, an instrument of social communication, a factor of standardization, a norm shared by all" (Jacob 1992). Paradoxically, this cartographic fragmentation leaves room for (or even provokes) certain informational alignments. A first level of alignment concerns meta-referentials such as technical standards (OGC, INSPIRE), map backgrounds that are imposed either by ease of use (Google Images) or by authorities (UN Clear Map), and cartographic projections (see Figure 6.8).

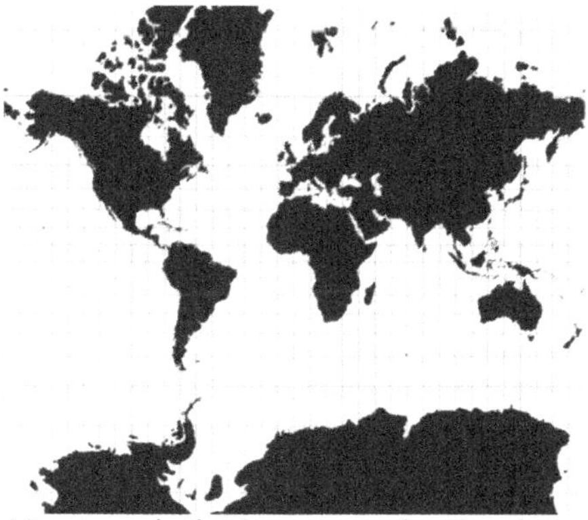

Mercator projection (preserves angles, strongly exaggerates the size of land near the poles), a *de facto* web mapping standard, used by Google Maps, Bing Maps and OpenStreetMap

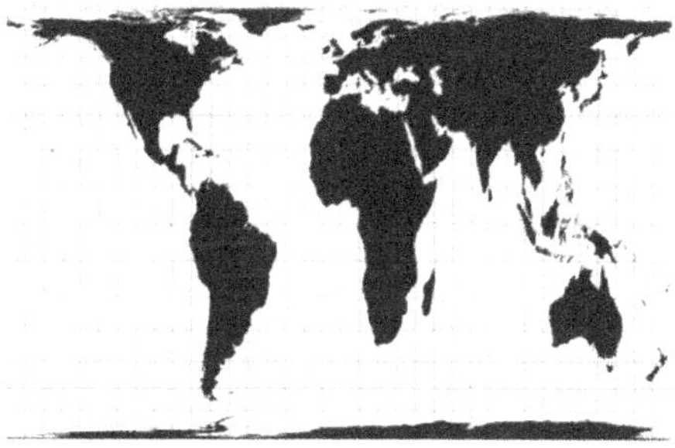

Gall–Peters projection (preserves surfaces but not angles) promoted by the United Nations in the 1980s

Figure 6.8. *World map projections*

The widespread use of the Mercator projection in Geoweb applications is shocking more than 40 years after Arno Peters' fight for a projection that preserves the size relationships between countries (Crampton 1994). Google defends the use of Mercator on the grounds that this projection maintains directions and that surface deformations are invisible at the scale generally used by the clients of its applications. The Mercator projection is therefore imposed on users of commercial mapping services (Google, Microsoft, Yahoo, etc.), for whom there are currently no alternatives. This example shows how Big Data mastered by companies can impose a specific form of representation on public organizations.

The second level of alignment concerns the nature of the information produced. Putting Geoweb data into a system creates an overlay of information that can both lead to fragmented, singular visions of the world, like the "little maps of the Web" (Noucher 2017), and allow for the common adoption of information that has demonstrated its broadest applicability (Bowker 2000, p. 662). Big Data and its algorithms can also lead to a reversal of the priority of concepts over measurement, even promoting a datacentrism that dispenses with theory (Anderson 2008). Big Data pushes observation to a level of detail beyond the dream of the 1:1 map (Jennings 2011), which raises the question of its generalization (Miller and Goodchild 2015, p. 455): Will the categories of quantification no longer be defined upstream of observation, according to conceptual frameworks established by science and politics, but downstream by algorithms, mastered by a reduced number of technicians?

6.6.3. *Political alignments*

With the emergence of new private actors in the processing of massive information, it seems necessary to go beyond the division between, on the one hand, political work on action objectives and, on the other hand, technical work on their statistical translation by supporting "seemingly technical discussions in the social debate open to the actors concerned" (Desrosières 2008, p. 71). The process of defining the SDG indicators, including various political and technical actors, seems to lend itself to such discussions. This is all the more true as such a framework can have a power of alignment (of methods, ways of thinking and political orientations), especially as it is not binding. This should not preclude articulation with binding quantification systems (defined for reporting under international agreements) or authoritative systems in a field (e.g. the Essential Climate Variables [ECV])[36].

Big Data are shaking up the current governance of statistics and geo-information; the future will tell whether it will continue along the path of multilateralism or whether the multinational information companies will push for agreements of an essentially commercial nature. More generally, the question of information justice arises in terms of technical, economic, legal and cognitive accessibility to Geoweb information, but also in terms of equitable spatial and thematic coverage of information. As a UN report points out, "many groups and individuals remain 'invisible', and many development issues are still poorly understood" (United Nations 2019). International statistical mapping, which aims to serve global environmental and development issues, depends on this changing information space, which is no longer in the hands of states or international organizations alone.

6.7. Conclusion

This chapter reports on the evolution of international statistical mapping since the 19th century, from the internationalization of statistical production in the inter-war period to the current context of Big Data. From the second half of the 20th century onwards, international organizations have been the main actors in international statistical cartography, which is still largely based on the collation of national data, but which is seeing increased possibilities of freeing itself from these data with the multiplication of data sources from remote sensing and Internet technologies.

36 A variable that contributes critically to the characterization of climate. There are currently 54 ECVs. A public consultation process allows for broad participation in the revision of the ECVs. Available at: https://gcos.wmo.int/en/essential-climate-variables/.

We have shown that international statistical mapping is now taking place in a global information space based on the Web, which is organized in infrastructures that allow the pooling and processing of increasingly diverse and massive data. We have seen that the arrival of Big Data and large digital companies such as GAFAM is challenging the hegemony of international organizations and states in the production of global carto-statistics. The example of the indicators of the SDGs has allowed us to illustrate the governance of a quantification of political objectives in the context of an international process.

In addition to questions of political and technical control of information, we have highlighted issues of informational and political alignment, cartographic visualization and informational justice. International statistical cartography will evolve very rapidly in the future, in view of the new partnerships between public and private actors in the field of geographic and statistical information, as well as the increasing availability of data. This will lead to fresh challenges in terms of comparability and permanence over time of quantifications made by a variety of actors and methods.

6.8. References

Anderson, C. (2008). The end of theory: The data deluge makes the scientific method obsolete [Online]. Available at: https://www.wired.com/2008/06/pb-theory/ [Accessed 28 February 2017].

Bowker, G.C. (2000). Biodiversity datadiversity. *Social Studies of Science*, 30(5), 643–683.

Buckland, M.K. (1991). Information as thing. *Journal of the American Society for Information Science*, 42(5), 351–360.

Crampton, J. (1994). Cartography's defining moment: The Peters projection controversy, 1974–1990. *Cartographica: The International Journal for Geographic Information and Geovisualization*, 31(4), 16–32.

Crampton, J. (2001). Maps as social constructions: Power, communication and visualization. *Progress in Human Geography*, 25(2), 235–252.

Cussó, R. (2012). L'activité statistique de l'Organisation économique et financière de la Société des Nations. Un nouveau lien entre pouvoir et quantification. *Histoire & mesure*, XXVII(2), 107–136.

Desrosières, A. (2000). L'histoire de la statistique comme genre : style d'écriture et usages sociaux. *Geneses*, 39(2), 121–137.

Desrosières, A. (2008). *L'argument statistique*. Presses de l'École des mines, Paris.

Etemad, B. (1994). Quelques grandes étapes de l'histoire de la collecte de statistiques rétrospectives. *Histoire & Mesure*, 91–100.

Facebook Connectivity Lab and Center for International Earth Science Information Network, CIESIN, Columbia University (2016). High resolution settlement layer (HRSL) [Online]. Available at: https://dataforgood.fb.com/tools/population-density-maps/ [Accessed 16 November 2020].

FAO (1948). Forest resources of the world. *Unasylva*, 2(4). Washington, DC [Online]. Available at: www.fao.org/docrep/x5345e/x5345e00.htm.

Fischer, F. (2012). VGI as big data: A new but delicate geographic data source. *GeoInformatics*, 3, 46–47.

Gautreau, P. and Noucher, M. (2013). Gouvernance informationnelle de l'environnement et partage en ligne des données publiques : politiques et pratiques de l'Open Data environnemental (Amérique du sud – France). *Netcom*, 27(1–2), 5–21.

Giuliani, G., Mazzetti, P., Santoro, M., Nativi, S., Van Bemmelen, J., Colangeli, G., Lehmann, A. (2020). Knowledge generation using satellite earth observations to support sustainable development goals (SDG): A use case on land degradation. *International Journal of Applied Earth Observation and Geoinformation*, 88, 1–12.

Goodchild, M.F. (2007). Citizens as sensors: The world of volunteered geography. *GeoJournal*, 69(4), 211–221.

Independent Group of Scientists appointed by the Secretary-General (2019). Global sustainable development report 2019: The future is now – Science for achieving sustainable development. Report, United Nations, New York [Online]. Available at: https://www.un-ilibrary.org/content/publication/5d04ad97-en.

Jacob, C. (1992). *L'empire des cartes : approche théorique de la cartographie à travers l'histoire*. Armand Michel, Paris.

Jennings, K. (2011). *Maphead*. Scribner, New York.

Joliveau, T. (2010). Le GéoWeb pour les nuls. *Monde géonumérique* [Online]. Available at: https://mondegeonumerique.wordpress.com/2010/06/24/le-geoweb-pour-les-nuls/ [Accessed 28 January 2020].

Kitchin, R. and Dodge, M. (2007). Rethinking maps. *Progress in Human Geography*, 31(3), 331–344.

Kraak, M.J., Ricker, B., Engelhardt, Y. (2018). Challenges of mapping sustainable development goals indicators data. *ISPRS International Journal of Geo-Information*, 7(12), 482.

Laney, D. (2001). 3D data management: Controlling data volume, velocity, and variety. Report, META Group Inc., Stamford [Online]. Available at: https://blogs.gartner.com/doug-laney/files/2012/01/ad949-3D-Data-Management-Controlling-Data-Volume-Velocity-and-Variety.pdf.

Lanly, J.-P. (1982). *Les ressources forestières tropicales.* Organisation des Nations unies pour l'Alimentation et l'Agriculture, Rome.

Le Goix, R., Ysebaert, R., Giraud, T., Lieury, M., Boulay, G., Louail, T., Ramasco, J.R., Mazzoli, M., Colet, P., Teurillat, T., Segessemann, A., Marcinczak, S., Bartosiewicz, B., Silva, E., Baerug, S., Holsen, T. (2019). ESPON Big Data for territorial analysis and housing dynamics. Wellbeing of European citizens regarding the affordability of housing. Report, ESPON, Luxembourg.

Lee, K. and Sener, I.N. (2020). Strava Metro data for bicycle monitoring: A literature review. *Transport Reviews*, 41(1), 27–47.

MacEachren, A.M. and Kraak, M.-J. (1997). Exploratory cartographic visualization: Advancing the agenda. *Computers & Geosciences*, 23(4), 335–343.

MacFeely, S. (2018). The Big (Data) Bang: What will it mean for compiling SDG indicators? Report, UNCTAD Research Paper, UNCTAD, Geneva.

Miller, H.J. and Goodchild, M.F. (2015). Data-driven geography. *GeoJournal*, 80(4), 449–461.

Noucher, M. (2017). *Les petites cartes du web : analyse critique des nouvelles fabriques cartographiques*. Éditions Rue d'Ulm, Paris.

Palsky, G. (1996). *Des chiffres et des cartes : naissance et développement de la cartographie quantitative française au XIXe siècle*. Comité des travaux historiques et scientifiques, Paris.

Pieroni, R. and Debarbieux, B. (2014). Données géographiques et construction institutionnelle d'une région transnationale. Le SOIA comme objet-frontière ? *Revue internationale de géomatique*, 24(1), 67–85.

Raffestin, C., Lopreno, D., Pasteur, Y. (1995). *Géopolitique et histoire*. Payot, Lausanne, Paris.

Reinsel, D., Gantz, J., Rydning, R. (2018). Data age 2025. The digitization of the world. From edge to core. Report, International Data Corporation (IDC), Framingham [Online]. Available at: https://www.seagate.com/files/www-content/our-story/trends/files/idc-seagate-dataage-whitepaper.pdf.

Sachs, J., Schmidt-Traub, G., Kroll, C., Lafortune, G., Fuller, G. (2019). Sustainable development report 2019. Report, Bertelsmann Stiftung and Sustainable Development Solutions Network (SDSN), New York.

Tobler, W.R. (1959). Automation and cartography. *Geographical Review*, 49(4), 526–534.

UN Global Pulse (2017). UN global pulse annual report 2017. Harnessing Big Data for development and humanitarian action. Report, UN Global Pulse, New York [Online]. Available at: https://www.unglobalpulse.org/document/un-global-pulse-2017-annual-report/.

UNECE High-Level Group for the Modernisation of Official Statistics (2013). Classification of types of Big Data. Conference of European Statisticians [Online]. Available at: https://statswiki.unece.org/display/bigdata/Big+Data+in+Official+Statistics.

UN/IEAG (2014). A world that counts. Mobilising the data revolution for sustainable development. Report, United Nations, New York [Online]. Available at: http://www.undatarevolution.org/.

United Nations (2019). The sustainable development goals report 2019. Report, United Nations, New York [Online]. Available at: https://unstats.un.org/sdgs/report/2019/ [Accessed 12 February 2020].

7

Indigenous Mapping: Reclaiming Territories, Decolonizing Knowledge

Irène HIRT
University of Geneva, Switzerland

7.1. Introduction

In the 1970s, the Indigenous peoples of the North and South of the Americas made their voices heard on the international scene to claim collective rights that they were denied at the national scale. The term "Indigenous" was coined to designate the peoples who felt they had not been able to achieve formal decolonization. In order to determine who is Indigenous, reference is generally made to the following criteria established by the United Nations (UN), and adopted by Indigenous peoples themselves:

– anteriority in a territory in relation to the colonial invasion;

– a non-dominant social or political position (regardless of demographic importance);

– the fact that a group considers itself distinct from the majority of society and wishes to preserve its collective identity and territory;

– self-identification (Martínez Cobo 1986, paragraph 379).

For a color version of all the figures in this chapter, see www.iste.co.uk/debarbieux/mapping.zip.

The Politics of Mapping,
coordinated by Bernard DEBARBIEUX and Irène HIRT. © ISTE Ltd 2022.

The historical relationship of Indigenous peoples and their lands, territories and resources is also considered to distinguish them from other cultural minorities (Daes 2001). Finally, the International Labour Organization (ILO) has extended the word "Indigenous" to groups marginalized because of their livelihood activities (hunting, gathering, slash-and-burn agriculture, etc.) in Asia and Africa, where the issue of ancestry is complex and controversial (Adivasi of India, San of Namibia, Batwa Pygmies of Central Africa, etc.). Today, a growing number of peoples undergoing forms of (neo)colonial domination consider being Indigenous as a political resource for defending rights that have been violated on a national scale. According to international organizations, 4–5% of the world's population considers itself or is considered Indigenous (between 370 and 500 million people), spread over 90 countries and five continents, and occupying 22% of the planet's land[1].

While it is impossible to speak of a single "Indigenous culture", common characteristics can be identified (Dobbs and Louis 2015, p. v). The same is true of cartographic knowledge and maps made *by* or *for* Indigenous peoples (hereafter "Indigenous mapping"[2]): contextual, cultural and historical differences do not preclude global trends. Overall, these cartographies refer to a dual reality that we will examine in turn in this chapter: on the one hand, they refer to Indigenous peoples' appropriation of the language and cartographic techniques of the modern state for the purposes of political emancipation and recognition of land or territorial rights from the 1960s–1970s onwards; on the other hand, they refer to knowledge traditions, forms of representation and cultural practices that are part of the ways of being-in-the-world of Indigenous peoples. Today, both types of processes, knowledge or practices coexist and interact with each other, mutually enriching each other in a dynamic perspective. Both are also profoundly affected by recent transformations in mapping and geographic information as a result of the rise of digital technologies, the Internet and multimedia.

1 The recognition of the rights of Indigenous peoples was enshrined in the *ILO Convention No. 169 Concerning Indigenous and Tribal Peoples* of 1989 (hereinafter ILO Convention No. 169) and in the *UN Declaration on the Rights of Indigenous Peoples* of 2007. ILO Convention No. 169 is the cornerstone of international law on Indigenous rights, since once ratified by states, it is binding, unlike the 2007 Declaration.

2 Different terms are used depending on the geographical or linguistic contexts or the approaches favored by cartographers: community-based mapping, "ethnocartography", *cartografia social* (in Spanish), etc.

7.2. Conceptual and theoretical milestones

The first part of this chapter proposes some conceptual and theoretical milestones, discussing the word "counter-mapping", which is frequently associated with Indigenous mapping. It then situates them in the field of academic thinking.

7.2.1. *Indigenous counter-mapping: a struggle for concrete and symbolic spaces*

> When our elders were blockading against the logging activities, the authorities and the people were destroying the forest. They told us to show them the maps of our territories and prove our claims. The Penan looked around and said, "This is our land!", only able to point at the surrounding area with their hands. But in their mind, they knew the exact traditional boundaries that their ancestors used a long time ago. The outsiders and the government did not understand the Penan culture nor did they recognize their land claims. When the Penan said, "This is our forest," they answered, "This is not Penan forest." That is when the Penan gathered together to brainstorm how to produce their own map of their land.

Komeok Joe, Penan community leader and mapping coordinator[3].

One of the major challenges of mapping for Indigenous peoples is to demonstrate their historical presence on their territories, and thus to be able to fight against the fiction of *terra nullius*, as illustrated by the statement quoted at the beginning of this section, from a video recounting the production of maps by the Penan of Malaysia since 2002. This work was supported technically and financially by the Bruno Manser Fund, and awarded the "Prix Carto 2019" by the *Société suisse de cartographie* (Swiss Cartographic Society). The Penan have identified more than 5,000 river names, named more than 1,000 other geographical objects (mountains, villages, etc.) and georeferenced more than 600 poisonous trees (see Figure 7.1). In recent decades, the absence of this type of information on official maps has allowed the state of Sarawak (Malaysia) to make the existence of the Penan and their occupation of the land invisible, and to grant concessions to logging and palm oil companies in areas presumed to be empty of inhabitants.

3 Video excerpt from the Bruno Manser Fund (Bruno Manser Fonds BMF 2018).

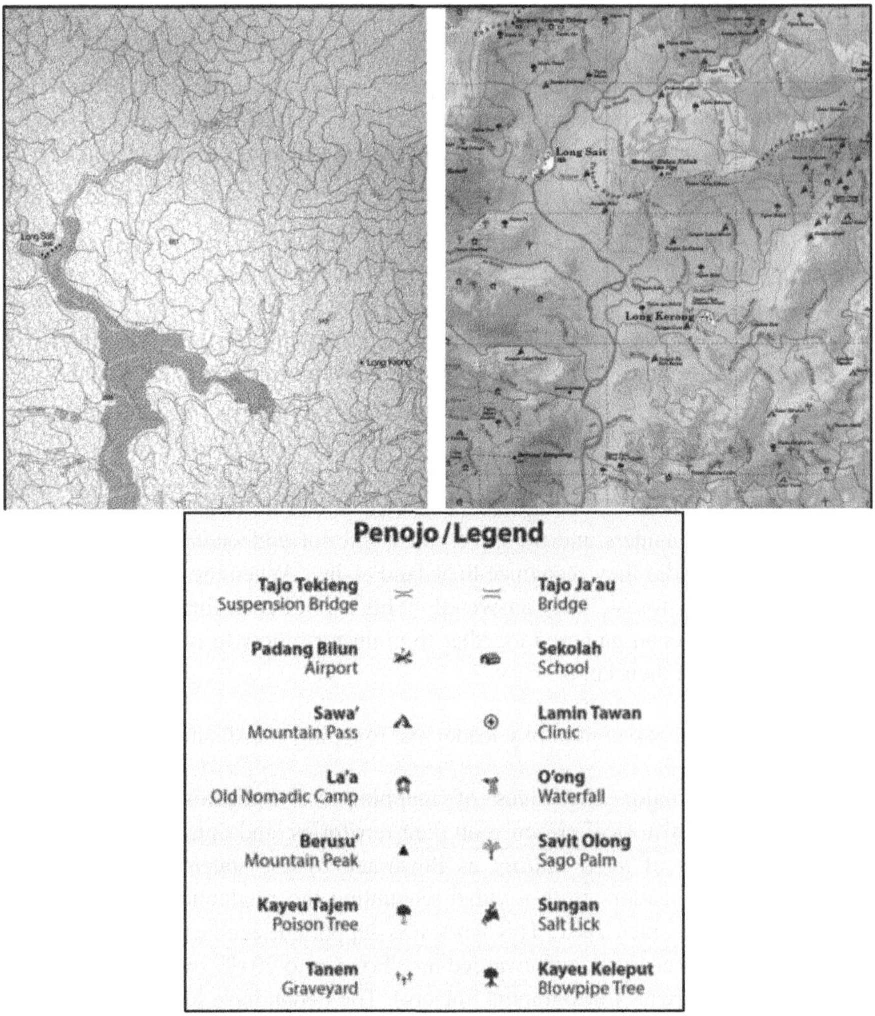

Figure 7.1. *Map of Penan territories before and after the inscription of Penan place names, with an extract of the legend (source: "Penan community maps for the rights of the Indigenous people of the Sarawak rainforest, Malaysia", Arcgis Storymaps, https://storymaps.arcgis.com/stories/7635a7357d264b558c46564f53443925, accessed November 13, 2020)*

This situation is far from unique. In the second half of the 20th century, many Indigenous peoples around the world became aware that their territories were not, or were poorly, represented on official maps, either because of a lack of knowledge or historical discrimination by governments, or to justify land grabbing (Herlihy and Knapp 2003).

In this context, Indigenous communities and organizations have appropriated the state's cartographic techniques and language in order to strengthen the legitimacy of their claims and to influence state-dominated land-use planning and land policies. Nancy Peluso describes the Dayak's production of alternative maps to those of the Indonesian state to defend their rights to territory and forest resources. She calls this strategy "counter-mapping" (Peluso 1995, p. 384). Other authors have spoken of "counter-hegemonic" or "counter-point" maps (Sparke 1998), "protest maps" (Wood *et al.* 2010) or "map insurrection" (Rundstrom 1991, p. 10). Peluso has compared the importance of this "vernacularization" of maps in the 20th century with the secularization of the language of print in the 15th and 16th centuries: both processes have made techniques for inscribing knowledge that was previously reserved for elites, accessible to the masses (Peluso 1995, p. 387). Counter-mapping is thus characterized by a political objective, namely the desire of Indigenous groups to challenge the exclusive power over maps of historically dominant actors (states, large corporations, etc.) and to increase their ability to represent themselves. According to Denis Wood, such radical attitudes take maps and mapmaking in a far more novel direction than new digital technologies do (Wood *et al.* 2010, p. 111). The word "counter-mapping" is now also associated with the struggles of other marginalized groups claiming sociospatial justice (see Kollektiv Orangotango (2018) for a range of such practices).

Indigenous peoples' counter-mapping practices aim to appropriate the tools and devices that have contributed to the dispossession of their lands, territories and resources over centuries of colonial rule. Their premise is summed up in Bernard Nietschmann's famous statement reversing Brian Harley's (1988, pp. 277–312) assertion that maps have been weapons of imperialism: "More Indigenous territory has been claimed by maps than by guns. This assertion has its corollary: more Indigenous territory can be reclaimed and defended by maps than by guns" (Nietschmann 1995). It echoes another well-known phrase repeated over and over in the literature: "Map or be mapped" (Stone 1998).

The political dimension of Indigenous counter-mapping practices is often reinforced by participatory mapping methodologies. In Indigenous contexts, these methodologies tend to emphasize the cross-cultural dimension of co-constructed knowledge by taking into account Indigenous epistemologies, ontologies and methodologies. For example, in some Mapuche communities in southern Chile, dreaming practices to communicate with the spirits of ancestors and those of the territory are common. In a mapping project carried out between 2004 and 2006 by the Mapuche-Williche[4] of Chodoy Lof Mapu, some participants thus considered dreams as a legitimate source of geographical information to map their territory, and that they complement other sources of knowledge (Hirt 2008, 2012). Furthermore, while counter-mapping is generally a collective process, it can also be an individual endeavor, as illustrated by the map of the Southern Cone of America in the 19th century, drawn in 2002 by Mapuche historian Pablo Marimán (see Figure 7.2).

This cartographic representation shows the territory from a Mapuche perspective. The map is oriented along an east-west axis according to the rivers, and shows the Andes not as an impassable barrier, but as a link between the two regions of *Wallmapu* (name given since the 1990s by the Mapuche to their ancestral territory, which extended on both sides of the Andes before the foundation of the Argentinian and Chilean States in the 19th century). The presence of the Mapuche in this territory is reasserted by place names in *Mapudungun* (Mapuche language). This representation is a counterpoint to the Chilean national imaginary which, since independence, has perceived the country along a north-south axis, reinforced by the construction of the country's main railway line and the Pan-American Highway. This example shows that counter-mapping as an alternative production to dominant maps does not only contribute to struggles for concrete spaces (land, territory), but also to struggles unfolding in the symbolic spaces of representations, imaginaries and knowledge production (Hirt 2017).

In this broad sense, Indigenous counter-mapping is challenging the "Great Divide" between Western science and Indigenous knowledge, and consequently also between modern scientific cartographies and Indigenous maps (Turnbull 2000). By "modern Western cartographies", we mean the scientific cartographic techniques, practices and representations developed from the European Renaissance onwards, which, in the words of Robert Rundstrom, have tended to impose a "Cartesian-Newtonian" and homogenizing vision of space (Rundstrom 1995, p. 45); the cartographic techniques and language of the state being only one way, among others, of practicing this modern cartography. However, as Annita Lucchesi rightly points out, Indigenous societies did not wait for scientific cartography or the modern state to produce maps. Historically, they have had their own knowledge and devices for understanding and codifying space (Lucchesi 2018). However, the recognition of these as "cartographies", along with modern Western cartography, was, until late in the 20th century, challenged by Western hegemony over the determination of what is considered legitimate and valid knowledge. For this recognition to occur fully, approaches arguing the symmetry of knowledge (Turnbull 2000) and inclusive definitions of mapping had to be developed, as we will see in the final part of this

4 The Williche are a regional sub-group of the Mapuche people.

chapter (Woodward and Lewis 1998; Turnbull 2000; Johnson *et al.* 2005; Edney 2018, pp. 68–79).

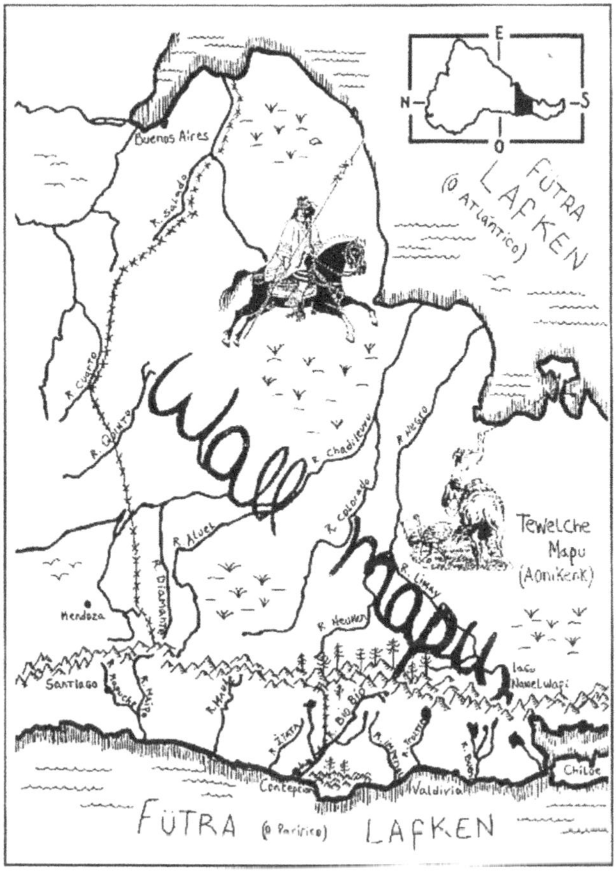

Figure 7.2. *Mapuche territory in the 19th century, drawn by Pablo Marimán (source: Marimán (2002, p. 54), reproduced with permission)*

7.2.2. *Indigenous mapping as a field of academic thinking*

Indigenous mapping was structured as an academic field in the 1990s (scientific publications prior to this date being relatively rare[5]), that is, three decades after the

5 The abundance of gray literature produced by non-governmental organizations (NGOs) or international organizations (manuals of "good practices", activity reports, etc.) is not inventoried here.

first contemporary production of maps by Indigenous peoples in the context of advocacy or development projects. For Rundstrom, this time lag between social practices and scholarly thinking can be explained by the fact that researchers have long considered mapping in Indigenous contexts as a practical exercise, limited to drawing up an inventory of successes, failures and working practices (Rundstrom 2009, p. 315). Furthermore, as they frequently work for NGOs and international cooperation agencies, publishing their work in an academic format is not usually a priority for them. It is therefore hardly surprising that the first special issue on Indigenous mapping (see Poole 1994) was published by *Cultural Survival*, an American Indigenous rights NGO.

Academic work on Indigenous mapping is at the intersection of critical cartography, geography, anthropology, development studies and, more marginally, law, the sociology of knowledge and post- or decolonial critique. Most of the publications are in English, and the most complete corpus concerns Canada, because of the pioneering use of counter-mapping by the Indigenous peoples of that country, as we shall see in the next section of this chapter. The publication of several special issues over the past 25 years on the subject (Poole 1994; Herlihy and Knapp 2003; Sletto 2009b; Louis *et al.* 2012; Dobbs and Louis 2015; Desbiens *et al.* 2020) is a sign of scientific vitality, although it remains a niche with a small community of researchers. Remarkably, since the 2000s, Indigenous geographers, mostly university-educated North American women, have entered the field of Indigenous mapping: Renee Pualani Louis of Hawaii (Louis 2004, 2007, 2017), Annita Lucchesi Hetoevėhotohke'e of the Cheyenne people (Lucchesi 2016, 2018), Mark Palmer of the Kiowa Tribe of Oklahoma (Palmer 2009, 2012), Margaret Wickens Pearce of the Citizen Potawatomi Nation (2008, 2014) and Stephanie Pyne of the Anishinaabe people (Pyne and Taylor 2012; Pyne 2013). Although still underrepresented in academics (Lucchesi 2018), these authors, all geographers, have contributed to the research on Indigenous mapping in three ways:

– first, by taking up the conceptual and cultural expansions of maps made by critical cartography in the 1980s and 1990s (see section 7.4);

– second, by drawing on Indigenous conceptions of cartography;

– third, by responding to Maori scholar Linda Tuhiwai Smith's (2002) invitation to decolonize knowledge and research methods.

In addition, cartography has been one of the areas used by the Indigenous Peoples Specialty Group of the American Association of Geographers to help Indigenous youth enter into academics (Louis *et al.* 2017).

7.3. Using maps to claim rights and reclaim territories

The second part of this chapter presents the history and the main debates related to the use of maps by Indigenous peoples in order to decolonize their relationship with the state and the actors of the dominant society. It also analyzes the limitations of the political use of maps.

7.3.1. *Brief history of the political use of maps*

Historians have shown that in Central America, the use of maps as a language of resistance to the colonizer, although marginal, was already a strategy of Indigenous peoples at the beginning of the European conquest (Harley 1992, p. 527). However, it was not until the 20th century that the use of maps by Indigenous peoples to defend rights to land, territory and natural resources became widespread. Initially sporadic in the 1950s in Alaska and Canada, it took off in Canada in the 1960s and 1970s before spreading in the rest of the world at the end of the 20th century.

The distribution of these practices across the globe, as well as the material conditions for their implementation and success, are uneven. In 2005, Chapin *et al.* identified three geographical regions:

– The first concerns Canada and Alaska and projects carried out with hunting, gathering, fishing and trapping societies, having developed a coherent set of methodologies and approaches.

– The second encompassed Asia, Africa and Latin America, and projects relating to both hunting and fishing societies and small farmers. Characterized by a diversity of terminologies and methodologies across countries or regions, the mapping projects of this second region are marked by development ideologies and goals.

– The third region coincides with the 48 US states south of Canada where, from the 1990s onwards, the Bureau of Indian Affairs (BIA) imposed a "top-down" mapping approach on Indian reservations, marked by advances in geospatial technologies but long lacking in participatory methodologies (Chapin *et al.* 2005, p. 622; Palmer 2009).

This regionalization, although still partially valid today, needs to be updated by including northern Europe (the Sami) and Oceania, especially Australia where maps are mobilized by Aboriginal peoples in their land and territorial claims (Chatwin 1988). However, a comprehensive inventory on a global scale has become an impossible task in the face of the proliferation of mapping projects, and due to the abundance of information about them, especially digital and online sources.

7.3.2. *Indigenous mapping in the Americas: similarities and differences between North and South*

The case of the two Americas deserves to be studied in greater depth, especially because this goes beyond North–South divisions. In Canada, the mapping of Indigenous territories was systematized in the 1970s by First Nations and Inuit, whose survival was threatened by the race to grab the still untapped resources of the North (gas, oil, hydroelectricity, mining resources). For these nomadic or semi-nomadic hunting, fishing and gathering societies, demonstrating their land use and occupancy became an urgent task. Following legal decisions in favor of the recognition of the "Aboriginal title"[6] the Canadian government introduced a new northern development policy, known as "comprehensive land claims". It is aimed at reducing legal uncertainty over territories claimed by Indigenous peoples and for which Indigenous rights had not been extinguished by treaty. In order to negotiate new treaties, the federal government financially supported Indigenous organizations from the second half of the 1970s onwards, to carry out studies (land use and occupancy studies) proving their historical continuity on their territories (Freeman 2011, p. 21; Bryan and Wood 2015, p. 57). The Inuit Land Use and Occupancy Project (ILUOP) (1976) is the most comprehensive of these studies. It was conducted by the Inuit Tapirisat of Canada, an organization founded in 1971 by Inuit leaders who believed "it was time to speak with a united voice on various issues concerning [the] development of the Canadian North and the preservation of Inuit culture" (Freeman 2019). The ILUOP aimed to document Inuit environmental practices and representations over a land and marine area of more than 2.8 million km^2 within what was then the Northwest Territories and northeast Yukon (Freeman 2011). Together with the Nunavut atlas (Canadian Circumpolar Institute *et al.* 1992), which specified the area that the Inuit would be able to concretely control, the ILUOP was used to support the negotiation of agreements leading to the creation of Nunavut in 1999, a new Inuit-governed administrative entity within Canada (Rundstrom 2009, p. 315).

The ILUOP inaugurated map biographies, an aggregation of individual hunting, fishing, gathering and trapping itineraries drawn on topographic maps, collected through oral testimonies (see Figure 7.3). These maps showed that far from being empty, as suggested by colonial imaginations, Inuit territories were full of place names, food caches, camp sites, and human and non-human migration routes. They have become a methodological standard in Canada (Chapin *et al.* 2005;

6 "Aboriginal title" refers to Indigenous rights recognized by the Royal Proclamation of 1763, but *de facto* ignored for a long time by the governments in Canada.

Freeman 2011) – framed by a manual (Tobias 2000, 2010) – and an inspiration for other Indigenous peoples around the world.

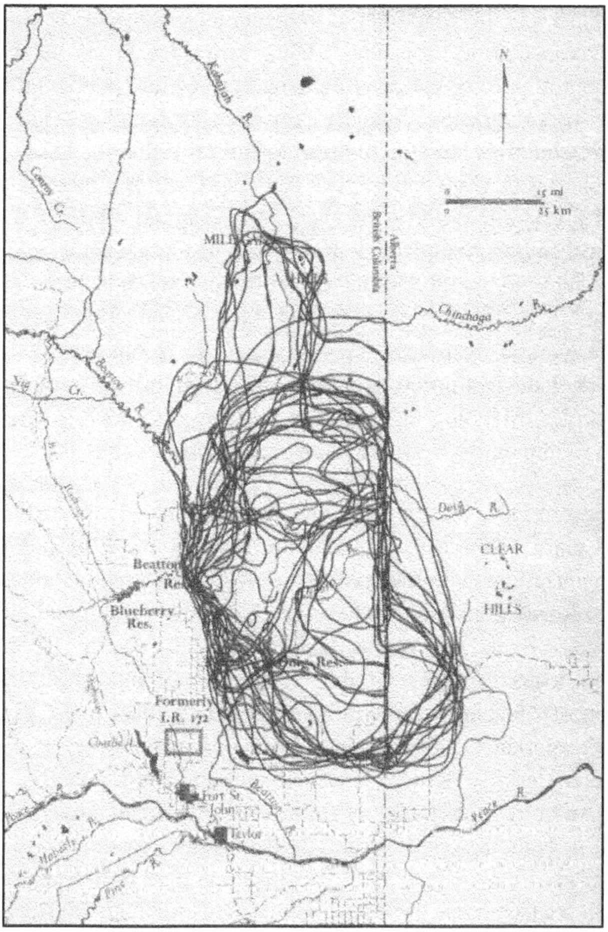

Figure 7.3. *Example of a map biography overlaying individual hunting routes (source: "Doig River Reserve: Hunting". Published by Faber & Faber. © H. Brody, 2002, p. 161. Reprinted with permission of the author c/o Rogers, Coleridge & White Ltd. in London)*

Anthropologist Hugh Brody helped bring these maps to a wider audience with his book *Maps and Dreams* (2002 [1981]). Alternating ethnographic accounts and research data collected in the context of a land use and occupancy study in the

1970s, Brody recounts the struggle of the Dunnez-za (Beaver) people against plans to build an oil and gas pipeline on their land.

In Latin America, the mapping of Indigenous lands and territories has been massively developed from the late 1980s onwards, in an explosive context combining recognition of rights, environmental protection, political decentralization and new areas of colonization linked to resource exploitation (Hale 2018). Many Latin American countries have recognized territorial rights for Indigenous peoples: first, by ratifying ILO Convention No. 169[7], and second, through constitutional recognition of the multicultural – or even multi-national (Bolivia and Ecuador) – characteristic of countries that thus broke with the two-century-old project of a homogeneous, mestizo nation. Some constitutional reforms have recognized the collective rights of Indigenous peoples (and Afro-descendants) over their ancestral lands and the natural resources they contain. In some cases, this has been accompanied by land restitution programs and even further agrarian reforms for Indigenous peoples. Mapping has not only been used to delimit newly recognized lands, but also to implement resource management plans.

Beyond these trends, the diversity of situations makes any exhaustive review difficult. In Central America, demarcation projects have been carried out by Indigenous organizations with the support of NGOs, research institutes and national or international foundations[8], thanks to research contracts and funding obtained from states or the World Bank. The World Bank has played a central role in conditioning the granting of loans to states on the recognition of Indigenous rights and the implementation of "development with identity" (Hale 2005, p. 17; Kymlicka 2013, p. 109). These experiences are examined in numerous publications by geographers (Joe Bryan, Peter Herlihy, Bernard Nietschmann) and anthropologists (Mac Chapin, Charles Hale, Anthony Stock), mainly from North American universities. They have been involved in Honduras and Nicaragua (Mosquito coastline), especially among the Misquitos, Garifunas, Pech and Tawahka, and in Belize, among the Maya (Chapin 1995; Nietschmann 1995; Chapin and Threlkheld 2001; Herlihy 2003; Stocks 2003, 2005; Bryan 2011; Hale 2018)[9].

7 15 Latin American countries (out of 23 worldwide) have ratified ILO Convention No. 169.

8 These include Cultural Survival, Native Lands, Right & Resources, Oxfam America and the World Wildlife Fund.

9 In 1992, Chapin *et al.* also produced a map with the National Geographic Society, showing the relationship between Indigenous peoples and the remaining forests in Central America, entitled *The Coexistence of Indigenous Peoples and the Natural Environment in Central America*, which can be found here: https://www.eli.org/sites/default/files/images/report-covers/10806_i1262-2.jpg [Accessed December 4th, 2021].

In several Latin American countries, the mapping of Indigenous lands and territories was developed by activists before being institutionalized or even state-run. In Bolivia, for example, at the end of the 1980s, Indigenous demands for the decentralization of power and the recognition of their own political and territorial organizations were heard by international cooperation actors. The latter became aware that giving priority to agrarian unions and state territorial divisions to the detriment of the Indigenous divisions would doom their development projects to failure. The European Union, Danish cooperation and the United Nations Food and Agriculture Organization (FAO) then financed the production of maps of the *ayllus* and *markas* (Aymara and Quechua territories). This initiative was the first time that the Indigenous peoples' views on territory were taken into account in Bolivia and, as a result, in the 1990s, Indigenous lands and territories (local forms of autonomy) gained legal recognition. This led to their cartographic demarcation throughout Bolivia (Hirt and Lerch 2013).

In contrast, Chile has remained outside these developments. Lacking constitutional recognition of Indigenous peoples and their territories, and despite Chile's belated ratification of ILO Convention No. 169, the mapping of Indigenous territories there still remains marginal[10].

Some ideas have been circulated between the North and South of the continent. Chapin, Nietschmann and Herlihy drew on the Inuit model for their projects in Central and South America (Wood *et al.* 2010, p. 132). Canada's mapping innovations were further disseminated to the South through the involvement of Indigenous leaders in international networks. Between 1975 and 1981, Secwepemc (Shuswap) leader George Manuel (1921–1989), president of the World Council of Indigenous Peoples, traveled to Argentina, Chile, Guatemala and Peru to promote land and occupancy studies as a means of struggling against state land grabs. Mapping ideas and strategies were probably also shared within the Working Group on Indigenous Peoples (WGIP), established in 1982 at the United Nations (Rundstrom 2009, p. 315). The WGIP brought together Indigenous delegates from around the world (including Manuel) for 25 years to draft the 2007 Declaration, serving as a melting pot for the development of a pan-Indigenous identity (Morin 2011). It can also be assumed that at the scale of the Americas, the existence of organizations such as the International Indian Treaty Council (IITC) may have fostered the circulation of ideas about mapping. The IITC, when it was founded in 1974 at Standing Rock, South Dakota, brought together representatives of 98 Indigenous nations from the Americas, the Caribbean and the Pacific.

10 See Hirt (2012) and Mansilla *et al.* (2019) for participatory mapping projects with Mapuche people's organizations or communities.

Finally, the advent of the Internet in the 1990s saw the proliferation of websites or networks for sharing mapping resources, methodologies or "best practices" online. The website "ppgis.net" focuses more broadly on participatory mapping methodologies or geographic information systems (GIS); others are specifically dedicated to Indigenous mapping, such as the Aboriginal Mapping Network (nativemaps.org). Moreover, others have held countless international conferences and workshops training Indigenous peoples in mapping tools.

7.3.3. *The aporias of the "cartographic-legal strategy"*

Indigenous counter-mapping practices have been celebrated as a form of revenge on colonial history and as an act of empowerment for Indigenous peoples. And undeniably, they are exactly this in several ways: they make Indigenous peoples and their territories visible, as in the earlier mentioned case of the Penan of Malaysia; they contribute to political processes that enable Indigenous groups to establish themselves as legitimate stakeholders and to sit at the negotiating table; and finally, they are part of the mechanisms and expertise that enable them to obtain financial compensation and resolve territorial conflicts. As the map by the Mapuche historian Marimán shows (see Figure 7.2), Indigenous cartographic representations also help challenge the hegemonic narrative of the origin and development of the state and the nation by conveying alternative meanings of space, territory and sovereignty (Sparke 1998). Finally, cartographic processes foster dynamics of cultural revitalization, providing opportunities for several generations to share knowledge and experiences, and to strengthen the sense of belonging to a collective (Éthier 2020, p. 46; Tobias 2000, pp. 1–2). And while these positive effects may not always be true, we have seen that, in many cases, Indigenous peoples must make maps if they want to make their existence and that of their lands and territories known and recognized.

Nevertheless, in a global context that is unfavorable to Indigenous peoples, maps reveal their double-edged nature. Peluso points out that counter-mapping is a "highly territorialized" strategy, since it takes place within states, which set the rules of the game according to their own legal and territorial principles (Peluso 1995, p. 394). Thus, the paradox of such a strategy is to go "against" the state while doing "the same" as it[11]. In other words, the "cartographic-legal strategy", to use Joel Wainwright and Joe Bryan's term, traps Indigenous peoples in the "state trap". In doing so, rather than reversing colonial relations, it may reconfigure them, reinscribing state power in Indigenous spaces, and leading Indigenous peoples to

11 I thank Pierre Lemonnier for this formula.

adopt the state's vision of territory and property rights (Wainwright and Bryan 2009, p. 167). Canada provides a good example of these contradictions: while the comprehensive land claims policy allows Indigenous peoples to produce their own maps, the state remains both judge and party, since land-use studies are funded by government loans. Moreover, for Indigenous peoples, proving the historical continuity of their presence on the territory is a headache when the spaces concerned have been destructured by colonization (Desbiens *et al.* 2015, pp. 191–208). They are thus confronted with a discrimination that is twofold: the denial of otherness, expressing the majority of society's refusal to recognize alternative forms of territoriality, and the "denial of coevalness" (Fabian 1984), through an injunction to authenticity which obliges Indigenous peoples to prove "traditional" ways of life that relegate them to a frozen past and "assign them to territoriality" (Hancock 2007, p. 73) in mostly rural spaces[12].

In Latin America, researchers question the ambivalence of the Latin American multiculturalist model (Gros and Dumoulin Kervran 2011). According to them, the success of the map depends on the effective application of constitutionally recognized Indigenous rights. The example of the 2001 Inter-American Commission on Human Rights (IACHR) ruling in favor of the collective land ownership rights of the Mayangna (Awas Tingni community, Nicaragua) is emblematic of these ambivalences; their victory remained symbolic for a long time, in the face of a state that did not feel obliged to implement the IACHR ruling, and in the face of a conservative bureaucracy, recalcitrant to initiate land demarcation (Hale 2005, 2018). And as in Canada, the map is seen in Latin America as part of a means of social pacification in resource-rich regions where the state wishes to promote economic investment, but which proves to be threatened by resources, land or territorial conflicts. From a state perspective, the legal and cartographic demarcation of Indigenous lands and territories would then function as a governmental rationality to instill a new sociospatial order under the guise of recognition (Bryan 2011, pp. 41–42).

In concrete terms, this reconfiguration of colonial relations obliges Indigenous peoples to translate their conceptions of space-time into that of the state (Thom 2014). This often amounts to a reformulation of a relationally structured socioterritorial organization – similar to an airline route map, with hubs and destinations – as a mosaic of discrete and contiguous entities (Thom 2009, p. 197). Thus, the cartographies accompanying the establishment of Indigenous property

12 Ironically, a study like ILUOP also serves Canada's geopolitical interests, mobilizing it to justify its sovereignty over Arctic lands and waters in the face of other nations' ambitions (Freeman 2011, p. 28).

rights do not so much reflect a past spatial order than they do reflect negotiated contemporary spaces (Bryan 2011, p. 40). One consequence of this is the profound transformation of group structures and identities, but also of relations between groups (Rundstrom 2009, pp. 314–318; Sletto 2009a). The act of social and spatial delimitation being deeply cultural (Thom 2009), it is particularly affected by this process of state "domestication" (Wainwright and Bryan 2009, p. 163). Moreover, belonging to more than one group and shared spaces between neighboring groups are particularly problematic. While the state generally ignores the former, it invariably views the latter as "overlappings" that Indigenous actors are required to resolve in order to achieve territorial entities controlled by a single group. According to Thom, the polygonal representation of state territory, based on stable residence (villages, "reserves", etc.), tends to break down broader territorial affiliations, particularly those of the kinship group (Thom 2009, 2014). Frequently, this process of land demarcation exacerbates already existing conflicts between neighboring or related groups, or creates new tensions (Chapin 1995; Fox 1998; Rundstrom 1998; Tobias 2000; Fox *et al.* 2005). In response to such problems, the Miskito of Honduras, perceiving the resolution of overlappings as impossible but tactically necessary to legally protect their lands, proposed a map that allowed for multiple possible readings: on the one hand, the communities drew "exclusive" lines in order to obtain property rights from the state; on the other hand, they signed agreements with each other to maintain ancestral practices of sharing territory and the reciprocal right to cross the mapped boundaries (Bryan 2011).

7.3.4. Controlling the flow of geographic information

While becoming visible is a decisive issue in proving the historical presence of an Indigenous peoples in a territory, it also makes mapped spaces more easily identifiable, and therefore potentially appropriable by third parties. This is particularly tricky in the case of natural resources that are necessary for the survival of Indigenous peoples, or of sites that have spiritual meaning for them but are coveted by third parties for their economic value. Controlling the flow of geographic information is therefore a crucial issue, made even more complex by the rise of digital technology, which facilitates the copying and distribution of data. For this reason, the decision to publish or to not publish maps is generally taken by researchers with the involved Indigenous communities (Birraux 2017). Producing and disseminating maps of Indigenous territories is not without risk, both for researchers and for Indigenous peoples, especially in countries where states consider these representations as subversive (Chapin *et al.* 2005, p. 620; Rundstrom 2009, pp. 314–318). Rundstrom reminds us that in Sarawak (Malaysia, Borneo) in the 1990s, a Penan with a map could be arrested. In such contexts, the alternative ("map or be mapped") is understood by Indigenous peoples according to a "cost-benefit" calculation (Rundstrom 2009, p. 317). Pierrette Birraux, for her part, recounts that when she mapped Yanomami territories in Brazil in the 1980s, she saw her work confiscated by Brazilian authorities (Birraux 2017). Some First Nations in Canada or tribes in the United states have their own economic and technical resources, and degrees of political-administrative autonomy that allow them to hire cartographers directly, or to have their own mapping service. They can then more easily control the potentially confidential dimension of the geographic information and maps they produce (Gagnon 2020; Éthier 2020). However, this is not the case for the majority of Indigenous peoples around the world, who depend economically and financially on the assistance of external actors to produce their maps.

The controversy surrounding *México Indígena* in 2009 was a reminder of the ethical and political issues surrounding geographic information in an Indigenous context, and its possible instrumentalization by third parties. This participatory mapping project, carried out by geographers and anthropologists from Kansas University (USA) and Oaxaca (Mexico), had the official objective of contributing to the empowerment of several Indigenous communities in Oaxaca through the demarcation and titling of their lands. Tacitly, however, it was also about collecting information on a region considered strategic by the US military (Bryan 2010). However, the Indigenous communities involved claimed that they had not been clearly informed that the funding was coming from the US military; they denounced what they saw as an attempt at "geopiracy", a neologism referring, by extension with biopiracy (Shiva 1997), to the collection and use of geographical information without the free, prior and informed consent of the main stakeholders (Ribeiro 2009; Bryan and Wood 2015; Hale 2018).

7.3.5. *The ambiguous role of the digital giants*

In conclusion, the rise of the Web giants since the 2000s appears to be transforming the evolution of relations between states and Indigenous peoples. Google Maps is becoming increasingly popular with Indigenous peoples, mainly because it is free to access its technology. In Canada, the company sponsors training workshops and partnerships with Indigenous peoples (CBC 2014). No critical work has yet systematically examined how these private actors are reconfiguring state–Indigenous relations: to what extent do digital giants contribute to the political empowerment of Indigenous peoples? In what ways do they promote new forms of colonialism (Pinto 2018)? There is no doubt that in the near future, the question of Indigenous peoples' sovereignty will no longer be posed only in territorial terms, but also in digital terms.

In June 2017, Google Maps partnered with Firelight, an Indigenous consulting[13] group, to add 3,000 Indigenous communities to its online map of Canada (CBC 2017). While this move is a first step toward decolonizing representations of territory by including Indigenous peoples on the Google Maps map of Canada, it only replicates the country's narrow, colonial geography (the "reserves" created in the 19th century in which Indigenous peoples were forcibly settled). Other maps would have to be drawn taking into account all the spaces occupied[14] or claimed by Indigenous peoples, as well as the spaces lived in and traveled by individuals. Today, the cartographic invisibility of Indigenous peoples in Canada also concerns their growing presence in urban spaces, the movement of people between these spaces and the reserves, and the continuity of trajectories and practices on ancestral lands. Cartographic representations have contributed to crystallizing the processes of social and spatial "reduction" (Simard 2003) caused by the creation of "reserves" in the 19th century. Can they now reverse these dynamics by showing the real extension and nature of contemporary Indigenous territorialities?

7.4. Decolonizing maps and cartography

So far, we have seen how Indigenous peoples have seized upon modern cartographic languages and tools to decolonize their relationship with the state. We will now examine how maps and cartography themselves have been decolonized at an epistemological level since the 1980s, in light of the revaluation of Indigenous peoples' own cartographic knowledge and the new possibilities offered by digital technologies and multimedia.

7.4.1. *Toward an inclusive definition of maps*

According to Matthew Edney, between 1830 and 1980, the history of cartography distinguished between Westerners producing "maps" and non-Westerners making "map-like objects". By 1980, a new history emerged that moved away from this ethnocentric viewpoint to focus on the diversity of "mapping modes" (Edney 2018, p. 72), that is, the cultural variations in the act of codifying

---

13 Firelight has been running an annual mapping workshop since 2014 (see : https://www.indigenousmaps.com/) training Indigenous communities in new digital and online tools (Google, Esri, QGIS, OpenStreetMap, etc.).

14 Other areas are administered or owned by First Nations and Inuit in Canada (e.g. treaty territories).

and representing spatial knowledge. This conception of history thus distanced itself from the colonial ideology according to which there would be "mapless" societies, in the same way as there would be "historyless" societies.

The cornerstone of this cultural decentering is the voluminous collection of books *The History of Cartography*, edited by David Woodward and later Edney, which Harley then contributed to[15]. One of the assumptions of the collection is that the cartographic impulse is both universal and particular (Harley 1987, pp. 1–42): what makes a map depends on the context, and the social and cultural needs it serves. Its functions are not limited to facilitating spatial orientation and wayfinding. They can also, among other things, highlight the prestige and power of the members of society who control its production and use it for political or religious ends (Woodward and Lewis 1998, p. 6)[16]. In the fourth volume, Woodward and Lewis distinguished three categories of mapping: mental or cognitive constructions; the oral translation of these mental constructions or their expression through performances and rituals, and finally, the materialization of this spatial knowledge in the form of an artifact (Woodward and Lewis 1998, p. 3).

This conceptualization has made it possible to consider artifacts that were previously ignored due to their non-conventional medium (from a Western perspective) as maps: ceramics, textiles, woodcuts, petroglyphs, bamboo and shell wefts, etc. From this perspective, the stick charts (Marshall Islands) or an Inca *khipu*, though with different forms and functions, are all ways of understanding and codifying space (Woodward and Lewis 1998, p. 1). This conceptual expansion has also allowed for the recognition of ephemeral forms of geographic inscription (e.g. drawings in sand or snow) as maps, and cartographic knowledge transmitted through oral language or performance (e.g. poetry, songs, dances, paintings, rituals, dreaming practices) (Woodward and Lewis 1998, p. 3). The songlines of the Aboriginal peoples in Australia thus constitute genuine spatial orientation and wayfinding devices (Chatwin 1987), as do the oral descriptions of hunting routes through the pack ice in Arctic communities (Aporta and Higgs 2005).

15 Carried by the Department of Geography at the University of Wisconsin-Madison, the project has six volumes, published between 1987 and 2015. The fourth, devoted to "African, American, Arctic, Australian and Pacific traditional societies" (Woodward and Lewis 1998), is the most interesting for this chapter.

16 This conceptualization is in dialog with non-representational approaches to mapping, seeing mapping as a set of social performances, practices and processes rather than as a final product (Kitchin and Dodge 2007).

7.4.2. *"Reconstructing" maps*

In the 1980s, the adoption of an inclusive definition of maps and mapping went hand in hand with the revaluation of Indigenous knowledge. In the following decade, researchers also examined the cultural effects of Indigenous peoples' mobilization of the tools, techniques and graphic language of modern cartographies.

Rundstrom launched the debate in 1995 in an article entitled "GIS, Indigenous Peoples, and Epistemological Diversity". He characterized GIS as a Western "techno-science" that profoundly transforms the Indigenous societies it reveals by (among other things) imposing a homogenizing and standardized vision of space (Rundstrom 1995, p. 45). This argument was then extended to geographic information technologies (GIT)[17]. This critique identified a series of antinomies between Indigenous knowledge and worldviews on the one hand, and GIT knowledge and worldviews on the other hand: for example, the opposition between "relational" Indigenous ontologies (LaDuke 1999) (characterized by continuities between humans and non-humans or the dead and the living) and the "naturalistic" ontology (Descola 2005) (which fragments and separates "nature" and "culture"), or the differences between oral knowledge systems, which are context-based and dynamic in opposition to inscribed and static knowledge, independent of their speakers, which Bruno Latour has designated as being "immutable mobiles" (Latour 2005 [1987]). Because of these differences, the translation of Indigenous knowledge and worldviews into the GIT's characteristic system of representation would run the risk of misinterpreting Indigenous geographies and assimilating them into modern cartographic epistemologies (Rundstrom 1995; Louis 2004).

In the 2000s, Indigenous cartographers both strong in their mastery of modern scientific cartographies and aware of the value and legitimacy of Indigenous mapping traditions sought to overcome these antinomies:

– Firstly, by linking oral tradition or performance-based forms of Indigenous mapping expressions with post-representational and performance-based approaches to critical cartography.

– Secondly, by showing that GITs are not *inherently* inappropriate for representing Indigenous knowledge and territorialities; on the contrary, they can be open to other uses and modalities of representation, and thus limit the possible misinterpretations of Indigenous geographies (we return to this point later, when we discuss the new potentialities of digital and online mapping and multimedia devices).

17 GIT encompasses GIS, web mapping services, remote sensing and GPS (Mericskay 2011).

– Thirdly, by seeking to move beyond the debilitating critique of Indigenous peoples' uses of modern cartographies to an approach that advocates the self-determination of knowledge.

In a major text, "Facing the future: Encouraging critical cartographic literacies in Indigenous communities", published in the journal *Acme* in 2005, Jay Johnson, Renee Louis, and Albertus Pramono acknowledged that modern scientific cartographies and state cartographic manners of representation are valuable in aiding their communities' overcome their struggles (described in the first part of this chapter), provided that they were used reflexively and with an awareness of their potential cultural assimilation effects (Johnson *et al.* 2005, pp. 91–94). Many Indigenous researchers have also argued that knowledge and techniques need to be adapted to suit different contexts, needs and objectives. For Annita Lucchesi, there is a need to reject injunctions of authenticity or identity assignments ("traditional" versus "modern", "Indigenous" versus "scientific", etc.) (Lucchesi 2018, pp. 22–23) in order to "reinvent" maps (Lucchesi 2018, p. 22). Others have emphasized the need to "reconstruct" them (Pyne and Taylor 2012, p. 93) or to create "hybrid spaces" (Palmer 2012) in the image of their own diverse identities.

7.4.3. *Representing "Indigenous depth of place"*

Over the past 15 years, these new perspectives have fostered a novel production of maps that seeks to creatively reveal Indigenous knowledge and territorial ontologies. Pearce and Michael Hermann have experimented with narrative and graphic techniques for including the multiplicity of experiences of place in maps (Pearce and Hermann 2010). As part of a reinterpretation of colonial maps in terms of "encounters" between the colonized and the colonizers (Harley 1992; Warhus 1997; Lewis 1998), both cartographers have traced, through ethnographic or historical sources and imagined information, Samuel de Champlain's travels in Canada, adding to the explorer's voice those of the Indigenous peoples who accompanied or met him (Pearce and Hermann 2010). In 2018, Louis published a book on Indigenous cartographic knowledge systems in Hawaii, describing performative practices such as *hula* (a Polynesian dance accompanied by songs from the Hawaiian Islands) as an encoding of spatial practices (Louis 2017). Finally, from a feminist perspective, Lucchesi (2016, 2018) has expressed, through maps and art, issues related to Native American genocide, sexual and domestic violence against Indigenous women (which she herself experienced), and the disappearances and murders of Indigenous women in Canada. Lucchesi has thus mobilized narrative cartography to imagine forms of restorative justice and to begin her own healing process (Lucchesi 2018, p. 24).

Much work has been inspired by what Pearce and Louis (2008) have called "Indigenous depth of place". Both cartographers used this term to show the dynamic, holistic and spiritual understanding of Indigenous space in Hawaii. Using GIS, they mapped the Ahupua'a of the island of Kaua'i[18] from a diachronic perspective, showing their evolution in relation to day and night light, the seasonal and tidal cycle, and the appearance of the Milky Way (Pearce and Louis 2008). Other researchers (both Indigenous and non-Indigenous) refer to an "Indigenous depth of place" to express a spiritual or time experience of a place, its invisible dimensions, memory of place, relations between tangible and intangible dimensions of territory and its ability to showcase Indigenous knowledge. Dreaming practices by Mapuche-Williche participants in a community-based mapping project in Chile are an example of this (see section 7.2.1) (Hirt 2008, 2012). With the aim of decolonizing territory and landscapes, geographer Justine Gagnon worked with the Innu of Pessamit in Quebec to represent their memory of place and their emotional connection to their territory, despite the latter's being flooded by the impounding of the Manic 5 dam in the 1970s (Gagnon and Desbiens 2018).

Interactive and online mapping has also made a significant contribution to renewing the cartographic representation and narration of Indigenous territorialities. The Geomatics and Cartographic Research Centre at Carleton University (Canada) has been developing cybercartographic atlases since 2003, led by D.R. Fraser Taylor, using open source software (http://nunaliit.org/) (Pyne and Taylor 2012). Four of these were produced with Inuit and First Nations people[19]. Their objectives, in addition to transmitting knowledge between generations, is to account for traditional Indigenous knowledge by linking together different forms of Indigenous expressive culture (oral traditions, material or visual culture, historical documents) (Caquard *et al.* 2009, p. 83). These atlases are also narrative tools for spatializing the history of colonial relations (e.g. relating to treaties and residential schools) for better mutual understanding between Indigenous and non-Indigenous peoples in North America (Pyne and Taylor 2012). Their interactive nature allows users to create their own content, and to determine how they wish to recount their relationship with the land and places. Taking advantage of the multimedia format, the atlases combine texts, photos, videos and audio with maps. In the *Kitikmeot Place Name Atlas*, each point on the map corresponds to a place name with an

18 Ahupua'a comprises land units divided according to resource distribution, ranging from the mountains to the ocean (Pearce and Louis 2008, p. 115).

19 *Atlas of Arctic Bay* (with Nunavut Arctic College); *Kitikmeot Place Name Atlas* (Kitikmeot Heritage Society); *Cybercartographic Atlas of Indigenous Perspectives and Knowledge of the Great Lakes Region* (Anishinaabeg communities), which includes a module on the 1850 Lake Huron Treaty.

associated audio file so that one can listen to its pronunciation in Inuktitut and/or a video of the Inuit speaker explaining the meaning of the place name in their own words. Figure 7.4 is a screenshot of the atlas examining the place name Ikpigyuaq. A video shows Inuit Elder Frank Analok describing Ikpigyuaq as "being a place where caribou were hunted in blinds with bow and arrow".

Figure 7.4. *Screenshot of the Kitikmeot Place Name Atlas (source: https://atlas.kitikmeotheritage.ca/, accessed November 13, 2020; reproduced with permission from the Geomatics and Cartographic Research Centre, Carleton University, and the Kitikmeot Heritage Society)*

It shows that the development of multimedia mapping has made it possible to overcome at least one of the criticisms voiced in the 1990s concerning the potential for cultural assimilation of modern Western cartography for Indigenous knowledge and worldviews. Indeed, such a multimedia device, by giving voice to Indigenous knowledge holders, makes it possible, on the one hand, to minimize the decontextualization of knowledge caused by the shift from oral to written forms, and, on the other hand, makes it possible to promote the direct expression, and therefore the most accurate possible, to promote the direct expression, and therefore the most accurate possible, of Indigenous worldviews and territorialities, thanks to the absence of an intermediary or translator (Pyne and Taylor 2012, p. 94).

However, Indigenous mapping in the era of interactive and contributive Web 2.0 is not spared from new factors of marginalization of Indigenous peoples. While there is not yet much work that has been done in this field, several authors examine the logics of exclusion of Indigenous knowledge in light of the universalization of contents and the constraints of data interoperability (Wellen and Sieber 2013; Noucher 2017; Reid and Sieber 2019). Matthieu Noucher traces the flow of *Teko* and *Wayãpi* place names (French Guiana), produced with participatory methodologies: from the Géoguyane portal to the *Geonames* global toponymic index, via open data portals located in France. At each stage, this dissemination has been synonymous with disassembly, loss and fragmentation of information, both because of the imperatives of interoperability and standardization, as well as the legal aspects related to open data. All of these factors have contributed to the misplacement and invisibilization of Indigenous place names "in the limbo of the Internet" (Noucher 2017, p. 56, Author's translation).

7.5. Conclusion

This chapter reflects the range of approaches and debates in the scientific literature on Indigenous mapping. We have seen that the latter term refers to two types of mapping processes: first, the contemporary appropriation of state cartographic language and techniques by Indigenous peoples to obtain recognition of their political, cultural, land and territorial rights. While these uses of maps by Indigenous organizations and communities are now well known and have been analyzed (particularly with regard to their effects in terms of empowerment or, on the contrary, the reinforcement of colonial relations), less is known about the consequences of the entry onto the scene of these new, extremely powerful private actors, the GAFA (Google, Amazon, Facebook, Apple). Their ability to reconfigure the relations between Indigenous peoples and states is yet to be understood. Second, we have shown that, in the broadest sense, Indigenous counter-mapping practices refer to the symbolic spaces of the decolonization of the imaginary, representations and knowledge. These practices contribute to a "new narrative" that (re)affirms the legitimacy of Indigenous cartographic knowledge and visions of the world and territory. It is in the digital and multimedia world that the future will undoubtedly hold the most surprises. This is all the more likely because the creative uses of these devices by a growing number of Indigenous cartographers run counter to any technological determinism. Clearly, as this chapter has argued, cartographic technologies do not *in themselves* lead to the cultural and political assimilation of Indigenous peoples, which is primarily caused by the imposition of the language and territorial principles of the modern state.

7.6. References

Aporta, C. and Higgs, E. (2005). Satellite culture: Global positioning systems, Inuit wayfinding, and the need for a new account of technology. *Current Anthropology*, 46(5), 729–753.

Birraux, P. (2017). From the Innu of Quebec to the United Nations, via the Yanomami of Brazil: An eventful journey of collaboration. *Justice spatiale/Spatial Justice JSSJ* [Online]. Available at: http://www.jssj.org.

Brody, H. (2002). *Maps and Dreams. Indians and the British Columbia Frontier.* Faber and Faber Limited, London [Original published 1981].

Bruno Manser Fonds BMF (2018). Penan Community Mapping: Putting the Penan on the map, July 24th [Online]. Available at: https://youtu.be/KwGdEzh1e3w.

Bryan, J. (2010). Force multipliers: Geography, militarism, and the Bowman expeditions. *Political Geography*, 29(8), 414–416.

Bryan, J. (2011). Walking the line: Participatory mapping, Indigenous rights, and neoliberalism. *Geoforum*, 42(1), 40–50.

Bryan, J. and Wood, D. (2015). *Weaponizing Maps: Indigenous Peoples and Counterinsurgency in the Americas.* The Guilford Press, New York.

Canadian Circumpolar Institute, Riewe, R.R., Tungavik Federation of Nunavut (1992). *Nunavut Atlas*. Canadian Circumpolar Institute and the Tungavik Federation of Nunavut, Edmonton, Alberta [Online]. Available at: https://www.uap.ualberta.ca/book-images/Open%20Access/9781772122183_WEB.pdf [Accessed 4 December 2021].

Caquard, S., Pyne, S., Igloliorte, H., Mierins, K., Hayes, A., Taylor, F. (2009). A "living" atlas for geospatial storytelling: The cybercartographic atlas of Indigenous perspectives and knowledge of the Great Lakes region. *Cartographica*, 44(2), 83–100.

CBC (2014). First Nations learn to map territories using Google Earth. *Canadian Broadcasting Corporation* [Online]. Available at: https://www.cbc.ca/news/aboriginal/first-nations-learn-to-map-territories-using-google-earth-1.2746110.

Chapin, M. (1995). Indigenous land use mapping in Central America. *Yale School of Forestry & Environmental Studies Bulletin Series*, (198), 195–209.

Chapin, M. and Threlkeld, B. (2001). *Indigenous Landscapes: A Study in Ethnocartography.* Center for the Support of Native Lands, Arlington, VA.

Chapin, M., Lamb, Z., Threlkeld, B. (2005). Mapping indigenous land. *Annual Review of Anthropology*, 34(1), 619–638.

Chatwin, B. (1988). *The Songlines*. Penguin Books, London.

Daes, E.-I. (2001). Indigenous peoples and their relationship to land. Final working paper prepared by the Special Rapporteur, Mrs. Erica-Irene A. Daes. Report E/CN.4/Sub.2/2001/21, United Nations, New York.

Desbiens, C., Hirt, I., Pekuakamiulnuatsh Takuhikan (2015). Développement industriel et négociations territoriales au Canada : défis et enjeux d'une nouvelle forme de traité. In *Terres, territoires, ressources : politiques, pratiques et droits des peuples autochtones*, Bellier, I. (ed.). L'Harmattan, Paris.

Desbiens, C., Hirt, I., Collignon, B. (2020). Indigenous spatial capital: Incorporating First Peoples' knowledges, places, and relations into mapping processes. *The Canadian Geographer/Le Géographe canadien*, 64(1), 4–64.

Descola, P. (2005). *Par-delà nature et culture*. Gallimard, Paris.

Dobbs, R. and Louis, P.L. (eds) (2015). Geospatial technologies and Indigenous communities engagement. Guest editorial preface. *International Journal of Applied Geospatial Research*, 6(1), iv–xiii.

Edney, M. (2018). Map history: Discourse and process. In *The Routledge Handbook of Mapping and Cartography*, Kent, A.J. and Vujakovic, P. (eds). Routledge, Milton Park, Abingdon.

Éthier, B. (2020). Analyzing entangled territorialities and Indigenous use of maps: Atikamekw Nehirowisiwok (Quebec, Canada) dynamics of territorial negotiations, frictions, and creativity. *The Canadian Geographer/Le Géographe canadien*, 64(1), 32–48.

Fabian, J. (1984). *Time and the Other: How Anthropology Makes its Object*. Columbia University Press, New York.

Fox, J. (1998). Mapping the commons: The social context of spatial information technologies. *The Common Property Resource Digest*, 45, 1–4.

Fox, J., Krisnawati, S., Hershock, P. (2005). *Mapping Communities. Ethics, Values, Practice*. East-West Center, Honolulu, Hawaii.

Freeman, M.M.R. (2011). Looking back – and looking ahead – 35 years after the Inuit land use and occupancy project. *The Canadian Geographer/Le Géographe canadien*, 55(1), 20–31.

Freeman, M.A. (2019). Inuit Tapiriit Kanatami (ITK). *The Canadian Encyclopedia* [Online]. Available at: https://www.thecanadianencyclopedia.ca/en/article/inuit-tapiriit-kanatami-itk.

Gagnon, J. (2020). De la carte au territoire : portée et usages des outils cartographiques au sein de la Première Nation des Pekuakamiulnuatsh Entrevue avec Michel Nepton, membre de la Première Nation des Pekuakamiulnuatsh et conseiller en aménagement du territoire. *The Canadian Geographer/Le Géographe canadien*, 64(1), 10–19.

Gagnon, J. and Desbiens, C. (2018). Mapping memories in a flooded landscape: A place reenactment project in Pessamit (Quebec). *Emotion, Space and Society*, 27, 39–51.

Gros, C. and Dumoulin Kervran, D. (2011). *Le multiculturalisme 'au concret' : un modèle latino-américain.* Presses Sorbonne nouvelle, Paris.

Hale, C.R. (2005). Neoliberal multiculturalism: The remaking of cultural rights and racial dominance in Central America. *PoLAR: Political and Legal Anthropology Review*, 28(1), 10–27.

Hale, C.R. (2018). Entre el mapeo participativo y la "geopiratería" : las contradicciones (a veces constructivas) de la antropología comprometida. In *Prácticas otras de conocimiento(s): Entre crisis, entre guerras. Tomo II*, Leyva, X., Alonso, J., Aída Hernández, R. (eds) [Online]. Available at: http://www.jstor.org/stable/10.2307/j.ctvn96g1f [Accessed 30 April 2020].

Hancock, C. (2007). "Délivrez-nous de l'exotisme" : quelques réflexions sur des impensés de la recherche géographique sur les Suds (et les Nords). *Autrepart*, 41(1), 69–81.

Harley, J.B. (1987). The map and the development of the history of cartography. In *The History of Cartography*, Harley, J.B. and Woodward, D. (eds). The University of Chicago Press, Chicago, IL.

Harley, J.B. (1988). Maps, knowledge, and power. In *The Iconography of Landscape: Essays on the Symbolic Representation, Design and Use of Past Environments*, Cosgrove, D. and Stephen, D. (eds). Cambridge University Press, Cambridge.

Harley, J.B. (1992). Rereading the maps of the Columbian encounter. *Annals of the Association of American Geographers*, 82(3), 522–536.

Herlihy, P.H. (2003). Participatory research mapping of Indigenous lands in Darién, Panama. *Human Organization*, 62(4), 315–331.

Herlihy, P.H. and Knapp, G. (2003). Maps of, by, and for the peoples of Latin America. *Human Organization*, 62(4), 303–314.

Hirt, I. (2008). Redistribuer les cartes : approche postcoloniale d'un processus de cartographie participative en territoire Mapuche (Chili). PhD Thesis, University of Geneva, Switzerland.

Hirt, I. (2012). Mapping dreams/dreaming maps: Bridging Indigenous and Western geographical knowledge. *Cartographica: The International Journal for Geographic Information and Geovisualization*, 47(2), 105–120.

Hirt, I. (2017). Enjeux de territoires, enjeux de savoirs. Jalons théoriques pour une interprétation transversale des reterritorialisations autochtones dans les Amériques. *Tsantsa. Revue de la Société Suisse d'Ethnologie*, 22, 112–122.

Hirt, I. and Lerch, L. (2013). Cartographier les territorialités indigènes dans les Andes boliviennes : enjeux politiques, défis méthodologiques. *Cybergéo. European Journal of Geography* [Online]. Available at: http://journals.openedition.org/cybergeo/25843.

Johnson, J.T., Louis, R.P., Pramono, A.H. (2005). Facing the future: Encouraging critical cartographic literacies in indigenous communities. *ACME: An International E-Journal for Critical Geographies*, 4(1), 80–98.

Kitchin, R. and Dodge, M. (2007). Rethinking maps. *Progress in Human Geography*, 31(3), 331–344.

Kollektiv Orangotango+ (ed.) (2018). *This is Not an Atlas: A Global Collection of Counter-cartographies*, 1st edition. Transcript Verlag, Bielefeld.

Kymlicka, W. (2013). Neoliberal multiculturalism? In *Social Resilience in the Neo-liberal Era*, Hall, P.A. and Lamont, M. (eds). Cambridge University Press, Cambridge.

LaDuke, W. (1999). *All Our Relations: Native Struggles for Land and Life*. South End Press, Cambridge.

Latour, B. (2005). *La science en action : introduction à la sociologie des sciences*. La Découverte/Poche, Paris [Original published 1987].

Lewis, G.M. (ed.) (1998). *Cartographic Encounters: Perspectives on Native American Mapmaking and Map Use*. University of Chicago Press, Chicago, IL.

Louis, R.P. (2004). Indigenous Hawaiian cartographer: In search of common ground. *Cartographic Perspectives*, 48, 7–23.

Louis, R.P. (2007). Can you hear us now? Voices from the margin: Using indigenous methodologies in geographic research. *Geographical Research*, 45(2), 130–139.

Louis, R.P. (2017). *Kanaka Hawai'i Cartography: Hula, Navigation, and Oratory*. Oregon State University Press, Corvallis, OR.

Louis, R.P., Johnson, J.T., Pramono, A.H. (2012). Introduction: Indigenous cartographies and counter-mapping. *Cartographica: The International Journal for Geographic Information and Geovisualization*, 47(2), 77–79.

Louis, R.P., Collignon, B., Hirt, I. (2017). Indigenous peoples and spatial justice. An interview with Renee Louis Pualani, translated by C. Cancel. *Justice spatiale/Spatial Justice JSSJ*, 11 [Online]. Available at: http://www.jssj.org/article/les-peuples-autochtones-et-la-justice-spatiale-un-entretien-avec-renee-louis-pualani/.

Lucchesi, A.H. (2016). –hóhta'hané: Mapping genocide & restorative justice in Native America. PhD Thesis, Washington State University, Pullman, WA.

Lucchesi, A.H. (2018). "Indians don't make maps": Indigenous cartographic traditions and innovations. *American Indian Culture and Research Journal*, 42(3), 11–26.

Mansilla, P., Melin, M., Royo, M. (2019). *Cartografía cultural del Wallmapu. Elementos para descolonizar el mapa en territorio mapuche*. LOM, Santiago.

Marimán, P. (2002). *Parlamento y Territorio Mapuche*. Instituto de Estudios Indígenas, UFRO/Escaparates, Temuco/Concepción.

Martínez Cobo, J. (1986). Study of the problem of discrimination against indigenous populations, Volume 5. Conclusions, proposals and recommendations. Report E/CN.4/Sub.2/1986/7/Add.4, United Nations, New York.

Mericskay, B. (2011). Les SIG et la cartographie à l'ère du géoweb : vers une nouvelle génération de SIG participatifs. *L'Espace géographique*, 40(2), 142–153.

Morin, F. (2011). Le malaise des anthropologues face à la globalisation de l'autochtonie. *Inditerra. Revue internationale sur l'autochtonie*, 3, 129–140.

Nietschmann, B. (1995). Defending the Miskito Reefs with maps and GPS: Mapping with sail, scuba, and satellite. *Cultural Survival Quarterly*, 18(4) [Online]. Available at: https://www.culturalsurvival.org/publications/cultural-survival-quarterly/defending-miskito-reefs-maps-and-gps-mapping-sail-scuba.

Noucher, M. (2017). *Les petites cartes du web. Approche critique des nouvelles fabriques cartographiques*. Éditions Rue d'Ulm/Presses de l'Ecole normale supérieure, Paris.

Palmer, M.H. (2009). Engaging with indigital geographic information networks. *Futures*, 41(1), 33–40.

Palmer, M.H. (2012). Theorizing indigital geographic information networks. *Cartographica: The International Journal for Geographic Information and Geovisualization*, 47(2), 80–91.

Pearce, M.W. (2008). Framing the days: Place and narrative in cartography. *Cartography and Geographic Information Science*, 35(1), 17–32.

Pearce, M.W. (2014). The last piece is you. *The Cartographic Journal*, 51(2), 107–122.

Pearce, M.W. and Hermann, M.J. (2010). Mapping Champlain's travels: Restorative techniques for historical cartography. *Cartographica: The International Journal for Geographic Information and Geovisualization*, 45(1), 32–46.

Pearce, M.W. and Louis, R.P. (2008). Mapping indigenous depth of place. *American Indian Culture and Research Journal*, 32(3), 107–126.

Peluso, N.L. (1995). Whose woods are these? Counter-mapping forest territories in Kalimantan, Indonesia. *Antipode*, 27(4), 383–406.

Pinto, R.A. (2018). Digital sovereignty or digital colonialism? *International Journal on Human Rights*, 15(27), 15–27.

Poole, P. (ed.) (1994). Geomatics: Who needs it? *Cultural Survival Quarterly Magazine*, 18(4) [Online]. Available at: https://www.culturalsurvival.org/publications/cultural-survival-quarterly/18-4-geomatics-who-needs-it [Accessed 4 December 2021].

Pyne, S. (2013). Sound of the drum, energy of the dance – Making the Lake Huron Treaty Atlas the Anishinaabe way. PhD Thesis, Carleton University, Ottawa.

Pyne, S. and Taylor, D.R.F. (2012). Mapping Indigenous perspectives in the making of the cybercartographic atlas of the Lake Huron Treaty relationship process: A performative approach in a reconciliation context. *Cartographica: The International Journal for Geographic Information and Geovisualization*, 47(2), 92–104.

Radio-Canada (2017). Google ajoute 3000 communautés autochtones à sa carte du Canada [Online]. Available at: https://ici.radio-canada.ca/nouvelle/1040852/google-carte-autochtones-premieres-nations-metis-inuit.

Reid, G. and Sieber, R. (2020). Do geospatial ontologies perpetuate Indigenous assimilation? *Progress in Human Geography*, 44(2), 216–234.

Ribeiro, S. (2009). Geopiratería en Oaxaca... y mucho más. *LaJornada*, January 31 [Online]. Available at: http://www.jornada.unam.mx.

Rundstrom, R. (1991). Mapping, postmodernism, Indigenous people and the changing direction of North American cartography. *Cartographica*, 28(2), 1–12.

Rundstrom, R. (1995). GIS, Indigenous peoples, and epistemological diversity. *Cartography and Geographic Information Systems*, 22(1), 45–57.

Rundstrom, R. (1998). Mapping, the white man's burden. *The Common Property Resource Digest*, 7–9.

Rundstrom, R. (2009). Counter-mapping. In *International Encyclopedia of Human Geography*, Kitchin, R. and Thrift, N. (eds). Elsevier, Amsterdam.

Sletto, B.I. (2009a). "Indigenous people don't have boundaries": Reborderings, fire management, and productions of authenticities in indigenous landscapes. *Cultural Geographies*, 16(2), 253–277.

Sletto, B.I. (2009b). Indigenous cartographies. *Cultural Geographies*, 16(2).

Shiva, V. (1997). *Biopiracy: The Plunder of Nature and Knowledge*. South End Press, Boston, MA.

Simard, J.-J. (2003). *La réduction : l'autochtone inventé et les Amérindiens d'aujourd'hui*. Septentrion, Quebec.

Smith, L.T. (2002). *Decolonizing Methodologies. Research and Indigenous Peoples*. Zed Books, London.

Sparke, M. (1998). A map that roared and an original atlas: Canada, cartography, and the narration of nation. *Annals of the Association of American Geographers*, 88(3), 463–495.

Stocks, A. (2003). Mapping dreams in Nicaragua's Bosawas reserve. *Human Organization*, 62(4), 344–356.

Stocks, A. (2005). Too much for too few: Problems of Indigenous land rights in Latin America. *Annual Review of Anthropology*, 34(1), 85–104.

Stone, M. (1998). Map or be mapped. *Whole Earth*, 54.

Tobias, T.N. (2000). *Chief Kerry's Moose. A Guidebook to Land Use and Occupancy Mapping.* Union of BC Indian Chiefs/Ecotrust Canada, Vancouver.

Tobias, T.N. (2010). *Living Proof: The Essential Data-Collection Guide for Indigenous Use-and-Occupancy Map Survey.* Union of BC Indian Chiefs/Ecotrust Canada, Vancouver.

Thom, B. (2009). The paradox of boundaries in Coast Salish territories. *Cultural Geographies*, 16(2), 179–205.

Thom, B. (2014). Reframing Indigenous territories: Private property, human rights and overlapping claims. *American Indian Culture and Research Journal*, 38(4), 3–28.

Turnbull, D. (2000). *Masons, Tricksters and Cartographers: Comparative Studies in the Sociology of Scientific and Indigenous Knowledge*. Routledge, New York.

Wainwright, J. and Bryan, J. (2009). Cartography, territory, property: Postcolonial reflections on indigenous counter-mapping in Nicaragua and Belize. *Cultural Geographies*, 16(2), 153–178.

Warhus, M. (1997). *Another America: Native American Maps and the History of Our Land.* St. Martin's Griffin, New York.

Wellen, C.C. and Sieber, R.E. (2013). Toward an inclusive semantic interoperability: The case of Cree hydrographic features. *International Journal of Geographical Information Science*, 27(1), 168–191.

Wood, D., Fels, J., Krygier, J. (2010). *Rethinking the Power of Maps.* Guilford Press, New York.

Woodward, D. and Lewis, G.M. (eds) (1998). *The History of Cartography, volume 2.3.* The University of Chicago Press, Chicago, IL.

8

Choramatic Representations in Participatory Processes of Territorial Projects

Sylvie LARDON
INRAE & AgroParisTech, UMR Territoires, Clermont-Ferrand, France

8.1. Introduction

In the current dynamics of territorial recomposition and globalization, new territorialities are emerging (Debarbieux and Vanier 2002), territorial projects figures are being developed (Debarbieux and Lardon 2003) and new engineering for sustainable development of territories is being implemented (Caron *et al.* 2017). In this context, the issue addressed here is that of the co-construction of territorial projects, bringing local actors together to enter into dialogue and mobilize public policies for collective action (Amblard *et al.* 2018).

In order to respond to this, within the framework of the Research-Training-Action programs at AgroParisTech – Centre de Clermont-Ferrand[1], we have

For a color version of all the figures in this chapter, see www.iste.co.uk/debarbieux/mapping.zip.

1 AgroParisTech – Centre de Clermont-Ferrand, France, develops training and research-action on the sustainable development of territories (see: http://www2.agroparistech.fr/Centre-de-Clermont-Ferrand-735.html). In the Research-Training-Action programs, training is designed as an interface between researchers and territorial development stakeholders and the driving force behind their interactions. The training mobilizes and generates new knowledge; it is based on and promotes the acquisition of skills.

designed a participatory prospective diagnosis approach: the “territory game” (Angeon and Lardon 2008; Lardon 2013). This approach aims to build a shared vision of the territory in order to develop collective action for territorial development, defined as an increase in the capacity of stakeholders to control the processes that concern them (Deffontaines *et al.* 2001, pp. 39–56).

The approach was tested for the first time at a conference in Clermont-Ferrand, France, jointly organized in 2003 by Debarbieux and Lardon (2003), on the figures of the territorial project as part of a prospective reflection by the French administration of territorial planning. At this conference, I argued in favor of the use of chorematic representations in this type of approach (Lardon 2003, pp. 109–129). Brunet (1986, p. 2) formalizes a choreme as “a choreme is an elementary structure of space, represented by a graphic model”, and that “any spatial configuration is the result of a possibly very complex combination of simple mechanisms”. At the same conference, Debarbieux (2003, p. 13) made the following observation in the form of a paradox: “territorial projects are full of images (but) iconography is very often the unthought of the project and of territorial prospective”.

Therefore, this chapter aims to reflect on the way in which choremes have been used to accompany the construction of territorial projects, as the link between the initiatives of local actors and the incentives of public policies, and on the way in which these choremes have been included in a participative prospective diagnosis process. To do this, I mobilize the nine issues of project iconography and territorial prospective, formalized by Debarbieux (2003, pp. 13–36):

– Three of these issues echo the methodological choices of such a diagnosis. Indeed, the iconography of the project and of the prospective has a political character (issue 3), due to the fact that this diagnosis concerns groups of people conducting territorial projects. As a result, it is necessary to produce synthetic images to minimize the flow of images (issue 1). The “experts” from outside the territory, traditionally mobilized in the form of an injection of knowledge into the internal reflection, are rather involved in the dialogue with the actors of the territory because of their expertise in territorial analysis (issue 5).

– In this chapter, I present three illustrations to show how the “territory game” approach produces results that meet three other challenges presented by project iconography. First, the choremes provide a comprehensive overview of a space, to make it intelligible and meaningful for the actors by differentiating the needs of iconography (issue 2) to communicate. Second, by relying on learning processes, they manage the distance between the territorial reality and its graphic representation (issue 4) by showing the different points of view of the actors. Third, by discussing the geographical framework, the distances and the objects that need to be represented, the actors arrange the images according to their needs (issue 6) to build the action.

– The last three issues mentioned by Debarbieux (2003, pp.13–36) provide solutions for interpreting and evaluating the "territory game" approach. What are the futures of the representations developed in a participatory manner (issue 8)? What place does esthetics occupy in these representations (issue 7) and what is the genericity or singularity of the representations (issue 9)? These issues open up a new role for iconography in the territorial development process.

Thus, three conditions for establishing territorial strategies respond to the political issues related to this process: the participation of stakeholders, the development of a common language and the inclusion in project approaches. With these conditions met, it is possible to articulate the implementation and appropriation of public policies with integrated and recognized development initiatives. These are the principles of the "territory game" approach.

8.2. Methodological choices

Developing a cross-sectional approach to spatial modeling and territorial development (Lardon *et al.* 2001) is a tool for dialogue between the various stakeholders concerned: territorial stakeholders, public policy stakeholders and research and training stakeholders (issue 3). The approach articulates research and action by mobilizing spatial representations for interaction between researchers and stakeholders around synthetic images (Caron and Cheylan 2005). The spatial modeling approach is formalized into a methodological itinerary that we set out "as a way of articulating different methods of processing information, but also of mobilizing actors, around the production of spatial representations, to progressively build a shared and strategic vision of the territory" (Lardon and Piveteau 2005, p. 77).

The "territory game" methodological itinerary allows for both an analytical and critical stance (issue 1). It takes place in three stages: a diagnosis, to identify the main structures and dynamics of the territory; a prospective, to imagine scenarios of evolution for the territory; and the formulation of courses of action, to respond to the issues identified in the territory. Prior to the "territory game", a diagnosis based on existing cartographic and statistical documents and surveys of key informers is used to construct the game sheets on cross-disciplinary themes. This also makes it possible, with the partners supporting the process, to identify the participants to be invited to the game, representative of these different themes, and of the types of actors involved locally (institutional actors, professionals and civil society). The knowledge of specialists and experts is thus combined with that of the inhabitants (issue 5). To facilitate interactions between actors during the game and their appropriation of the approach, plenary feedback sessions are held between production, as well as table discussions in smaller groups. Afterwards, a public feedback stage allows the players to see what was produced during the game, enriched by the researchers' analyses of the new ideas brought up at the collective table discussions. This step back allows for reconstituting the entire spatial reasoning and giving meaning to the approach, by explaining it step by step. All of the actors concerned with the territory, present at or receiving the restitution, are thus involved.

8.3. Three illustrations of the use of choremes in territorial projects

Developing a common language enables the various local and institutional stakeholders to understand one another, to articulate the scales and integrate the issues, to link the local and the global and to implement public policies locally. The approach consists of systematically translating the information and knowledge produced at each stage of the methodological itinerary, in the form of chorematic representations. It is a powerful tool for synthesizing dynamics, comparing points of view and moving from one level of organization to another for better territorial integration. The drawings made by the actors are interpreted through the prism of the organizing principles of space (Lardon and Piveteau 2005) in order to have a more generic language that can be understood by all: the interlocutors who brought the information, the participants in the game, and the actors to whom they are communicated. This double translation process makes choremes intermediate objects, according to Vinck (1999), in the process of collective appropriation. The following three illustrations, taken from the 60 or so approaches carried out since the initial formalization of the approach at the 2003 conference, show the three main properties of choremes for communicating, integrating and acting.

8.3.1. *The Pikogan forest in Quebec or choremes for communicating*

In 2010, Quebec adopted a law on sustainable forest management[2], aiming to involve local populations in resource management by asking them to set up

2 Sustainable Forest Development Act, Quebec, 2010. http://legisquebec.gouv.qc.ca/fr/ShowDoc/cs/a-18.1.

community forest projects[3]. In 2012, the opportunity arose for a team made up of French students – as part of the international module of AgroParisTech's ACTERRA specialized Master's degree[4], – and Quebec auditors from UQAT[5] in continuing education, co-supervised by French and Quebec teachers, to support the community of Pikogan, part of the Algonquin First Nations of Abitibi-Témiscaminque, in the development of their community forest project.

We spent 2 weeks in the field. During the first week, we conducted surveys with key informers for the forest management diagnosis. We participated in the activities of the community members by going with them into the forest where they talked to us about each tree and each hut, and by traveling with them in a canoe along the river. This immersion in the territory allowed us to better understand the way of life for the community and the meaning of their activities[6]. This facilitated communication, despite the problems arising from language and culture differences. During the second week, we worked with them to develop the community forest project, based on the current dynamics and what they needed to value in their activities, to produce a shared vision for the future of the forest and identify the development actions to be implemented. In addition, we held a session at the school to have the children draw their vision of the Pikogan forest. The final report gave an account of what the students had understood about the territory, its dynamics and the projects of the actors. At the same time, Carlos Kistabish, an artist from the community, produced a painting expressing the vision that we shared collectively.

The chorematic representations drawn up by the students to reflect the main dynamics of the territory were the basis for discussions with the community members, providing a simplified overview of what is important to them (see Figure 8.1). For the development of scenarios and the outline of courses of action, the students had prepared a panel of actions, represented by (symbolic) icons that the participants could choose or invent, in order to focus the exchanges on the story of activities in the

3 Community forests consist of the delegation of the management of public forest lands and certain resources to local and Indigenous communities. Their purpose is to promote the socioeconomic development of these communities by giving them greater decision-making powers and responsibilities regarding the development of a territory and some of its resources. Definition taken from the website of the Ministère des Ressources naturelles: http://www.mrn.gouv.qc.ca/forets/gestion/evolution-nouveau-regime-forets-proximite.jsp.

4 http://www2.agroparistech.fr/-MS-ACTERRA-Action-publique-pour-le-developpement-durable-des-territoires-et-de-.html.

5 Desjardins Chair in Small Community Development: https://www.uqat.ca/recherche/chaire-desjardins/.

6 See the video montage of these activities: http://www2.agroparistech.fr/podcast/Flash-international-du-Mastere-Acterra-Quebec-mai-2012.

territory. The children's drawings and the artist's painting echoed the sensitive dimension of the experience and helped to make the implicit explicit (Olmedo 2016, pp. 179–186).

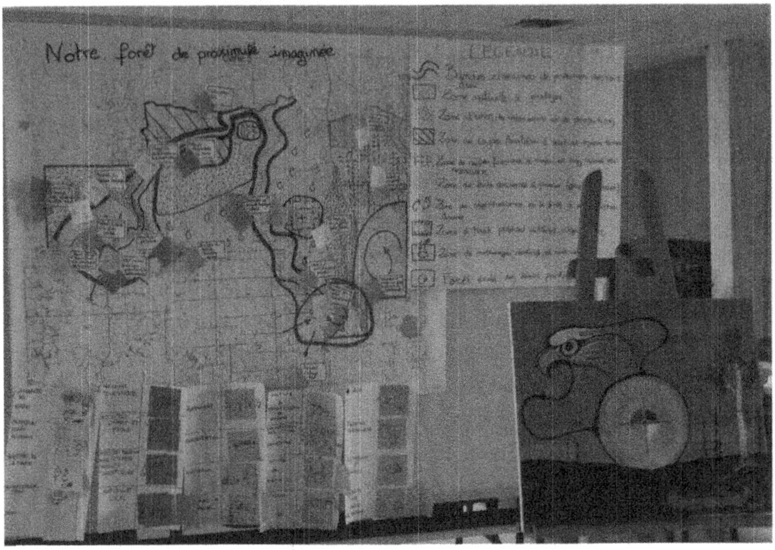

Figure 8.1. *The Pikogan Proximity Forest: chorematic representation, action icons and painting by Carlos Kistabish (© Lardon 2012)*

Thus, the participation of stakeholders, from managers to school children, was ensured in particular by the use of a variety of representation methods (maps, drawings, icons, choremes, tables) that made it possible to build a common language. The actors found tools adapted to their challenges and their capacity for action. Unfortunately, the project was not completed as the continuity of Quebec's federal forest policy was not guaranteed (Chiasson and Beaulieu 2020). We were not able put the spatial representations produced and exchanged between local actors to the test against political actors.

8.3.2. *The Western Pacific Islands or the choremes for integrating the issues*

In 2014, a former AgroParisTech auditor (continuing education on territory diagnosis), working in New Caledonia at the *Agence des Aires Marines Protégées* (Marine Protected Areas Agency), asked us to assist her in her project to build a shared vision of the territory of the Western Pacific, prior to the extension of new protected areas. This project was part of the European PACIOCEA program[7], which aims to identify and enable sustainable management of resources for local populations that are highly dependent on marine ecosystems. With two other colleagues from AgroParisTech and the Nouméa Marine Protected Areas Agency, we co-hosted a collective session with international researchers specializing in environmental issues (the area being described as a biodiversity hotspot) and local representatives from each of the 22 islands with various institutional statuses (independent or not).

The session took place over three intensive days of co-production. The game sheets had been produced beforehand by the Marine Protected Areas Agency based on a typology of ecosystem services. While maintaining the quality of the data for an audience of scientific or regional experts and the rigor of the itinerary, the exercise left room for creativity and the expression of the tangible and intangible values held by the players. The collective productions exceeded expectations, despite the difficulties inherent to the specificity of the situation: a marine territory, actors who are specialists in the subject, a common foreign language (English), significant environmental and social issues, as well as a heterogeneous facilitation team.

The table of spatial organizing principles made it possible to express and analyze the chorematic representations of the diagnoses, issues and scenarios of the territory's evolution, as produced by the different groups, based on the information provided and the knowledge exchanged. The scenarios revealed the strategic choices of the participants, such as the "Pacific Union" scenario (see Figure 8.2) addressing the issues of resilience and food self-sufficiency through intra-regional cooperation (Littaye *et al.* 2016). The reconstruction of the underlying reasoning, conducted *a posteriori* by the researchers, reveals a very strong capacity for actors to integrate social and environmental concerns into a holistic vision. Thus, the elements of the diagnosis, which were strongly focused on climate change, gradually acquired meaning because they affected the population and its activities and encouraged the stakeholders to project themselves by providing socioeconomic responses to the ecological emergency and, in doing so, to regain control of their future. The approach was continued by a participatory workshop in the same way, with representatives of local governments with the same level of involvement as the actors.

7 The PACIOCEA program was launched in 2013 to analyze and cross-reference environmental, socioeconomic and cultural issues in order to improve the management of the marine environment on a large scale. In particular, it aimed to identify and enable sustainable management of resources for local populations that are highly dependent on marine ecosystems.

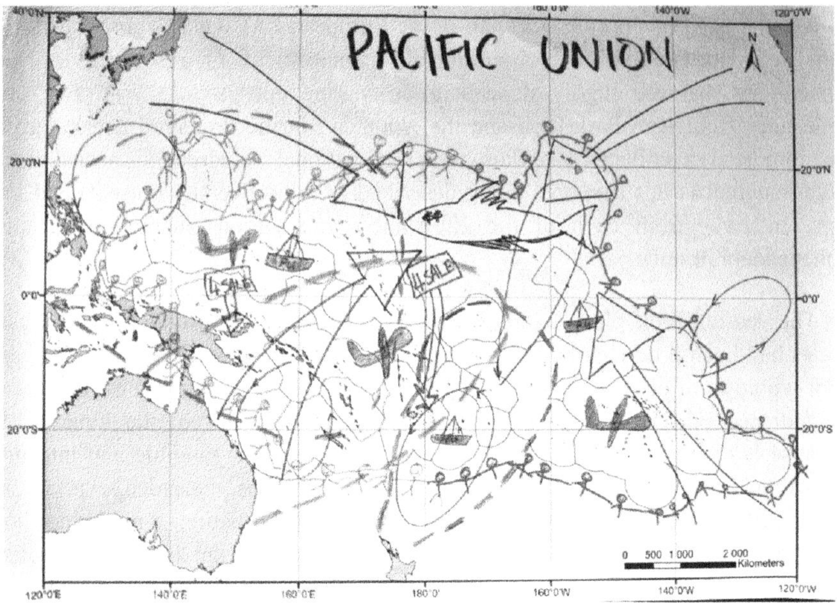

Legend Scenario
« Pacific Union »

N°	Symbol	Description
1		Less influence, more partnership with Australia
2		Less influence, more partnership with New Zealand
3		China influence receeding
4		Return migration whereas in pacific opportunities
5		Income from export of fishing resources
6		Pacific owned and operated fishing fleets
7		Strengthened collaboration at all levels and Pacific sub-groups
8		Long term resettlement plan
9		Patrol drones

Figure 8.2. *The Pacific Union scenario from one of the expert panels (source: Littaye* et al. *2016)*

This capacity for integration was supported and accompanied by the use of chorematic representations in the participatory process, both with international experts and with local governments. The action-research project was completed, but we do not know whether the actions were implemented by regional policy across the entire Western Pacific.

8.3.3. *The gardens of the Billom area in France or the choremes for acting*

Within the framework of the INVENTER project – *Inventons nos territoires de demain* (Inventing our territories of tomorrow) (2015–2020)[8] – we have conducted several "territory games" on the theme of food, from the beginning of the local declination of this public policy[9]. The territory concerned was the project territory of the *Pôle d'Equilibre Territorial et Rural du Grand Clermont* (GC) and that of the *Parc Naturel Régional Livradois-Forez* (PNRLF) (France). It was awarded the *Projet Alimentaire Territorial* (PAT)[10] label by the *Programme National pour l'Alimentation* in 2017.

In 2016, the game on the scale of the GC and PNRLF territory made it possible to show that food was not only one of the issues of the territory, but also that it was unifying other issues such as the environment, health and even the repopulation of town centers. Despite contrasting territorial dynamics between the GC, a rather urban territory, and the PNRLF, a rather rural territory, local actors projected the same development model, including food supply chains and distribution of processing and distribution sites (Lardon and Vergnaud 2019). In 2017, the game at the scale of the Billom Communauté territory, a transition area between the GC territory and the PNRLF territory, had fulfilled this vision and initiated a reflection on vegetable gardens as a place of food autonomy for many inhabitants of the territory (Lardon *et al.* 2020).

In 2019, the participatory workshop conducted with students of the Master of Geography at Clermont-Ferrand highlighted that the garden was an integrating spatial object, not only as a place of production for food self-sufficiency at lower cost, but also for exchanging practices that respect the environment and are adapted

8 See www6.inrae.fr>psdr-inventer.

9 Law of the Future of Agriculture, Food and Forestry of 13 October, 2014. See: https://www.gouvernement.fr/action/la-loi-d-avenir-pour-l-agriculture-l-alimentation-et-la-foret

10 See www.legrandclermont.com/projet-alimentaire-territorial-pat-du-grand-clermont-et-du-pnr-livradois-forez.

to climate change, and also as a place of sociability and cultural exchanges. Alongside the *Régie de Territoire des deux Rives*[11], this led us to create the gardeners' network. Located in Billom, this network enabled us to collect their points of view on the gardens, the gardening practices and the gardeners' motivations. In order to deepen the multicultural dimension of the gardens, participative workshops were conducted in 2020 that were specifically directed toward the elderly and people with disabilities, highlighting a shared sensitivity to the garden object and a common potential producer of meaning.

In the first "territory game" in 2016, the objective was to create a project dynamic between local actors and to legitimize the technical (GC and PNRLF) and scientific partners involved in the PSDR INVENTER project, as mediators, at the interface between public policy and local initiatives. Having received the PAT label in 2017, the political dimension was prioritized by the technical actors and the researchers focused their work on the food practices of inhabitants (Trimech and Lardon 2021, pp. 279–301). Thus, in the "territory game" in 2017, it was the actors' drawings, collected and synthesized by the students, which were used. The chorematic synthesis of the actor drawings of the commune of Billom highlights its role as a central source of organic and local products and cites the gardens as a place of food supply (see Figure 8.3). In the participatory workshops with gardeners in 2019 and 2020, a variety of media were used as follows: maps, photographs, word clouds, sound recordings (for the visually impaired), drawings, etc., to facilitate the expression of the participants and the identification of objects that had meaning for them.

This system currently allows for exchanges between gardeners and the structures that support them, to integrate social, environmental, cultural and economic issues, supported by the technical and political actors of the PAT and to implement concrete actions that are both locally anchored in the practices of the inhabitants and are also echoed by regional and national public bodies.

Thus, all the conditions were met in this project for action to be taken: participation of the actors, taking into account the diversity of their visions and political support by local elected officials. But it was also the emergence of new hybrid actors who were capable of articulating scales and worlds, such as the *Régie de Territoire des deux Rives*, that was highlighted. Research has also played an important role in this process of co-construction of territorial projects.

11 http://regiedes2rives.fr/.

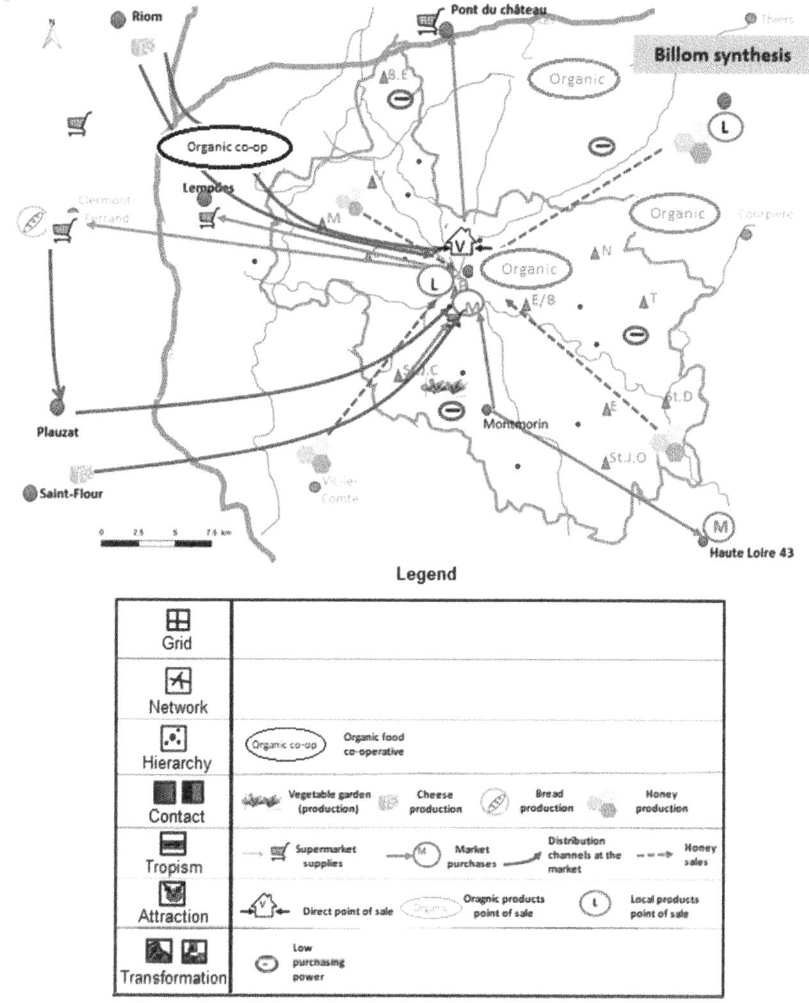

Figure 8.3. *Chorematic synthesis of drawings made by actors in Billom in 2017 (source: Trimech and Lardon 2021)*

8.4. How can we evaluate this co-construction of territories?

These examples illustrate how it is possible to build development models for rural and peri-urban territories based on the dynamics of actors involved in a territorial project. Choremes also have the properties of anticipation and creativity and open up a generic approach.

With regard to future representation (issue 8), it is increasingly clear that territorial prospective benefits from being taught by and with spatial representations, including being recognized by political actors. Thus, Vanier and Lajarge (2008) have shown in their work for the *Délégation à l'Aménagement du Territoire et à l'Action Régionale* that reflection on peri-urban futures, based on widely disseminated spatial models, is gaining in depth. Here, chorematic representations are used to express the actors' projections of the future by feeding their reasoning. Esthetics are not absent from these representations (issue 7). Without being metaphorical, these images can touch the "reader" more than a map with overly rigid standards! The actors can adhere to the proposals, or at least understand and feel them. This facilitates their appropriation by translating, through sensitive iconographies, the logic of the actors, in their singularity and their genericity (issue 9).

In the end, the choremes bring three properties to the territorial development process. First of all, they allow the participation of all stakeholders, including civil society. Spatial representations make visible these social innovations, which themselves contribute to territorial development, as suggested by Klein and Harrisson (2006), or Moulaert and Nussbaumer (2008). Second, the choremes formalize the main structures and dynamics of the territory in the form of spatial models, while explaining the logic of spatial reasoning; there is a collective development of spatial knowledge, similar to the situations studied by Le Ber (2008). Finally, the constructed approach is rigorous, reproducible and creative, flexible and securing the common interest, as in the cases of the collective and pictorial building sites conducted by Magnaghi (2014); it generates – other than by the norm and the rule – the capacity for collective action for territorial development. This brings an understanding of the world, integrative and creative.

The reflections that took place during the experiments of the "territory game" – which confronted us with public policies, their appropriation by the local actors, but also their recognition by the institutional actors – allow us to rethink the support for territorial development, in particular by the research. This leads to reconsidering the concepts and methods of territorial engineering, by reversing the logics. It is no longer just a matter of building project territories but territories of life, following the proposal of Latour (2017). It is no longer just a matter of matching institutional incentives and local initiatives, between top-down and bottom-up approaches, but of generating meaning and making it visible at the global scale, and building visions and sharing them at the local level, including their political dimensions (Lardon 2020). However, while the production of spatial representations has played a decisive role in helping local actors to understand one another and to express the same project for their living territory, they still struggle to be a language recognized by politicians. This is a challenge for future research in order to support territorial development dynamics.

8.5. References

Amblard, L., Berthomé, K., Houdart, M., Lardon, S. (2018). L'action collective dans les territoires : questions structurantes et fronts de recherche. *Géographie economie société*, 2(20), 227–246.

Angeon, V. and Lardon, S. (2008). Participation and governance in territorial development projects: The "territory game" as a local leadership system. *International Journal of Sustainable Development*, 11(2/3/4), 262–281.

Brunet, R. (1986). La carte-modèle et les chorèmes. *Mappemonde*, 86(4), 2–6.

Caron, P. and Cheylan, J.-P. (2005). Donner sens à l'information géographique pour accompagner les projets de territoire : cartes et représentations spatiales comme supports d'itinéraires croisés. *Géocarrefour*, 80(2), 111–122.

Caron, P., Valette, E., Wassenaar, T., d'Eeckenbrugge, G.C., Papazian, V. (2017). *Des territoires vivants pour transformer le monde*. Quae, Versailles.

Chiasson, G. and Beaulieu, H. (2020). Staple state et participation des communautés rurales dans la gouvernance de la forêt publique québécoise : bilan des mesures des réformes du régime forestier. *Revue organisations & territoires*, 29(1), 99–108.

Debarbieux, B. (2003). Neuf enjeux de l'iconographie de projet et de prospective de territoire. In *Les figures du projet territorial*, Debarbieux, B., Lardon, S. (eds). Éditions de l'Aube/DATAR, La Tour d'Aigues.

Debarbieux, B. and Lardon, S. (eds) (2003). *Les figures du projet territorial*. Éditions de l'Aube/DATAR, La Tour d'Aigues.

Debarbieux, B. and Vanier, M. (eds) (2002). *Ces territorialités qui se dessinent*. Éditions de l'Aube/DATAR, La Tour d'Aigues.

Deffontaines, J.-P., Marcelpoil, E., Moquay, P. (2001). Le développement territorial : une diversité d'interprétations. In *Représentations spatiales et développement territorial*, Lardon, S., Maurel, P., Piveteau, V. (eds). Hermes Science Publications, Paris.

Klein, J.-L. and Harrisson, D. (eds) (2006). *L'innovation sociale : émergence et effets sur la transformation des sociétés*. Presses de l'Université du Québec, Quebec.

Lardon, S. (2003). Diagnostic de territoire et représentations spatiales : les chorèmes, graphes et jeux. In *Les figures du projet territorial*, Debarbieux, B., Lardon, S. (eds). Éditions de l'Aube/DATAR, La Tour d'Aigues.

Lardon, S. (2013). Developing a territorial project: The "territory game", a coordination tool for local stakeholders. *FaçSADe*, 38, 1–4.

Lardon, S. (2020). Les trois vi-es des territoires ruraux : voyage en France ou ailleurs. *Revue Organisations & Territoires*, 29(1), 149–163.

Lardon, S. and Piveteau, V. (2005). Méthodologie de diagnostic pour le projet de territoire : une approche par les modèles spatiaux. *Géocarrefour*, 80(2), 75–90.

Lardon, S. and Vergnaud, D. (2019). Le projet inventer : la construction des territoires de demain. *Horizons Publics*, special issue Autumn 2019, 44–49.

Lardon, S., Maurel, P., Piveteau, V. (2001). *Représentations spatiales et développement territorial*. Hermès Science Publications, Paris.

Lardon, S., Beauseroy, O., Lalanne, L. (2020). Les jardins potagers, lieux d'échange et d'apprentissage collectif. *Reflets et perspectives de la vie économique*, LVIII(1), 35–48.

Latour, B. (2017). *Où atterrir ? Comment s'orienter en politique*. La Découverte, Paris.

Le Ber, F. (2008). De la bergerie au centre de calcul : élaboration collective de connaissances spatiales. *Revue d'anthropologie des connaissances*, 2(2), 143–145.

Littaye, A., Lardon, S., Alloncle, N. (2016). Stakeholders' collective drawing reveals significant differences in the vision of marine spatial planning of the western tropical Pacific. *Ocean & Coastal Management*, 130, 260–276.

Magnaghi, A. (2014). *La biorégion urbaine. Petit traité sur le territoire bien commun*. Eterotopia, Paris.

Moulaert, F. and Nussbaumer, J. (2008). *La logique sociale du développement territorial*. Presses de l'Université du Québec, Quebec.

Olmedo, É. (2016). Pour une cartographie affective des récits de femmes de Sidi Youssef Ben Ali (Marrakech, Maroc). In *Cartographier les récits*, Fournier, M. (ed.). CERAMAC, Clermont-Ferrand.

Trimech, A. and Lardon, S. (2021). Les cartes à dires d'acteurs et les chorèmes comme outils de diagnostic partagé des pratiques d'approvisionnement alimentaire dans le territoire. In *Cartographie des parcours*, Fournier, M and Troin, F. (eds). Presses Universitaires Blaise Pascal, Clermont-Ferrand.

Vanier, M. and Lajarge, R. (2008). Les futurs périurbains de la France en Europe. Travaux en ligne N°1 [Online]. Available at: http://periurbain.cget.gouv.fr/sites/default/files/les%20futurs%20p%C3%A9riurbains%20de%20la%20France.pdf.

Vinck, D. (1999). *Ingénieurs au quotidien. Ethnographie de l'activité de conception et d'innovation*. Presses Universitaires de Grenoble, Grenoble.

9

National Territorial Prospective Maps

Anne BAILLY
Honorary Technical Director, AEBK,
Maps and Communication and Consultant, France

9.1. Introduction

Some time ago, Claire Bretécher, a famous comic strip artist, portrayed the discord between a conquistador and a cartographer in the 15th century. Going against all documentation, he wanted her to make the coastline more complex and to add a number of islands to enhance his conquests and for it to "look pretty"! (see Figure 9.1).

As a cartographer, also trained in geography, working in support of the definition of regional planning policies, this story made me laugh a lot, and it made me question my professional practice. In the 1990s, I had the opportunity and the honor to work for the French administration in charge of national land use planning, then called DATAR (*Délégation à l'Aménagement du Territoire et à l'Action Régionale*). The orders included the production of prospective maps, that is, the design of images of the territory of France in the future.

9.2. DATAR, foresight and mapping

DATAR, created in the 1960s, was a prestigious agency of the French State, with a long-term vision for the evolution of the national space, in particular of the

---

For a color version of all the figures in this chapter, see www.iste.co.uk/debarbieux/mapping.zip.

The Politics of Mapping,
coordinated by Bernard DEBARBIEUX and Irène HIRT. © ISTE Ltd 2022.

development and balance that should characterize it. It was its departments that initiated major industrial development projects, such as the industrial port complexes of Fos-sur-Mer and Dunkirk, tourism projects, such as those on the Languedoc coast, and energy projects, such as the increase in hydroelectric power generation along major rivers, as well as nuclear power.

Figure 9.1. *Islands, the cartographer's workshop (excerpt). Maps and figures of the Earth. © Claire Bretécher/Hyphen-comics Paris 2021 (source: Centre Georges Pompidou (1980, p. 469))*

Compared to the administrations of other European countries, DATAR stood out because of its high level of interest in territorial foresight, although it varied over the decades. The term may have been used to describe different practices over time, but the aim at least was to always promote ways of thinking about the future of the French territory in order to guide public action. It was also a matter of communicating this future plan to convince the French of the validity of planning policy or to open up the debate.

Paradoxically, for a department dedicated to the territory, DATAR did not have a mapping department in the early 1990s; it called on external service providers to meet its needs. In fact, its large scale but localized projects did not require national land mapping since other organizations, such as the *Institut Géographique National* (French National Geographic Institute) or the Land Registry Service, performed this function.

Independent cartographers were few in number at the time and still worked largely manually, mainly in publishing. In the planning sector, despite the importance of the needs related to the production of "master plans for urban development and planning" on the scale of the main urban areas and regions, there we were even fewer. However, during the 1990s, in parallel with the increase in demand that led to the creation of cartographic departments in many administrations, the situation really changed with the development of geographic information systems, as well as the introduction of desktop publishing (DTP) in cartography; one increased the capacity to make maps 10-fold while the other offered new graphic perspectives.

9.3. The Paris Basin White Paper

My first intervention for DATAR took place in 1991 for the mapping of the White Paper on the Paris Basin; it conditioned all my subsequent interventions. As part of a renewal of territorial foresight, DATAR had launched a series of national thematic (transport, economy, city, rural) and territorial projects, the work being carried out at the scale of seven major inter-regional projects. The Paris Basin, encompassing seven of the 22 regions at the time, was one of them.

The study was carried out with the DATAR team in charge of this project, in collaboration with the decentralized services of the state: the SGAR (*Secrétariat Général pour les Affaires Régionales des préfectures de region,* General Secretariat for Regional Affairs of the regional prefectures). They were provided with a set of cartographic specifications, designed by myself, as well as base maps, the objective being that the seven regions would be speaking the same language. The synthesis work was done in collaboration with the DATAR study team, cartography being one of the important elements of the presentation.

The White Paper presents and analyzes the territorial problems of this macro-region in an exhaustive manner, supported by statistics and maps. It then proposes three prospective scenarios: one, the so-called trend scenario, entitled the "concentrated metropolis" in which the Paris conurbation plays a central role and captures most of the urban growth in a vast area that neutralizes the peripheral regions; another intermediate distribution scenario, entitled the "centralized multipolar metropolis", which identifies 10 satellite hubs whose growth would make it possible to limit the exponential expansion of the Paris conurbation and ensure the development of regional metropolises; a third, more proactive scenario of dynamization called the "connected metropolitan network", is betting on a new organization that coordinates the major urban centers into urban systems of European size and competence (see Figure 9.2). This third scenario transforms the 10 regional metropolises of the intermediate scenario into three metropolitan areas able to combine their complementarities and to develop their competitiveness together to counterbalance the strong polarization of Paris.

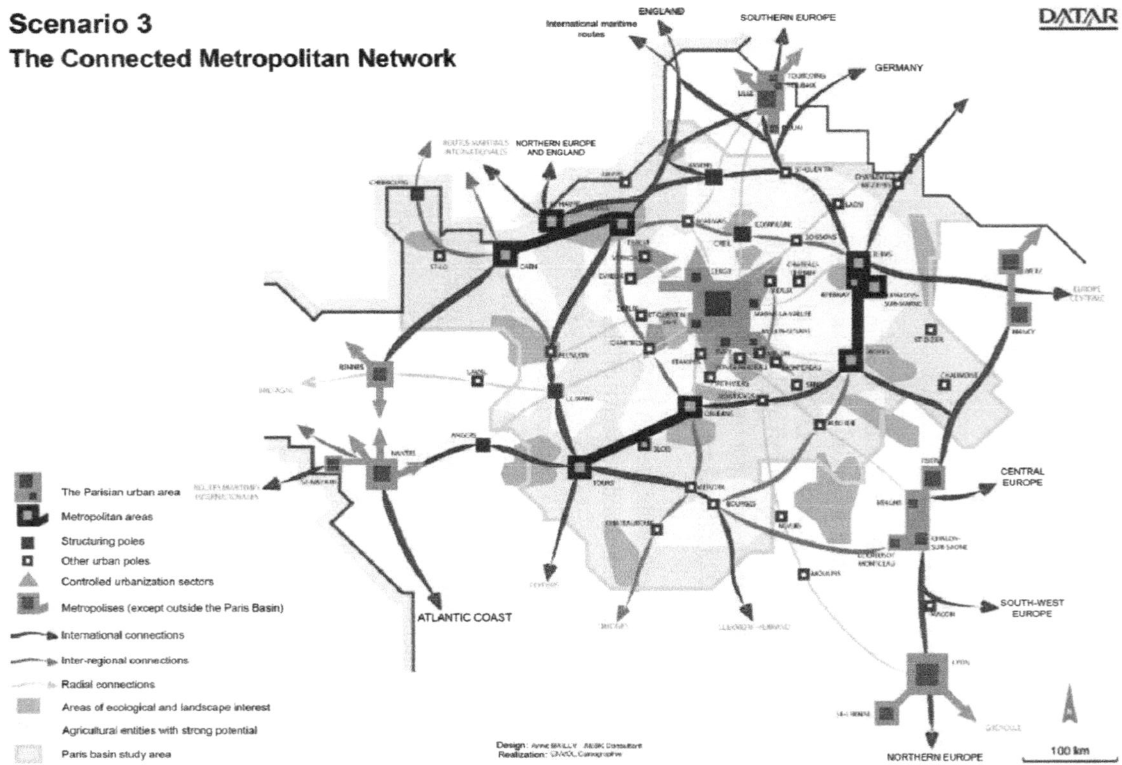

Figure 9.2. *Scenario 3 of the Paris Basin (source: DATAR (1992, p. 61))*

9.3.1. *Cartographic design*

As this was a prospective project, the maps produced were not intended to be a representation of geographical space of the type that prevails in maps concerned with accurate measurements and precise locations, such as topographical maps. Indeed, the function of prospective maps is rather to draw the general ideas of a structure or a territorial equilibrium, without focusing on the faithful representation of forms prevalent in scientific cartography. For this reason, I have opted for a deliberately simplified and very "graphic" mode of cartographic representation, similar to diagrams: while the base map used for the entire White Paper positioned the administrative boundaries, the main towns, the major communication routes and the structure of the hydrographic network, the boundaries were drawn at right angles or 45°, the towns were reduced to squares of different sizes, and the routes were related to simplified segments.

9.3.2. *The graphic vocabulary*

Every map adopts its own graphic vocabulary, playing on shapes, lines and colors. This vocabulary benefits from being inspired by the principles of graphic semiology which, like those set out by Jacques Bertin in a work written more than 50 years ago and never equaled since (1967), guarantee optimal map legibility. However, the vocabulary of prospective maps, because of their specificity, must be distinguished from that adopted for maps that are guided more by a concern for accuracy. In the case of the Paris Basin White Paper, it was necessary to translate all the political intentions of each of the three contrasting scenarios. Over the years, I have continuously developed my own "library" of signs and symbols that can be adapted to suit needs. It is not analogous to any graphic grammar intended to reflect the shape of existing objects within a geographical space, as in the case of topographic maps, or the structural elements of this space as identified by geographical analysis, as in Roger Brunet's chorematics (1986). All of my patterns are geometric (square, circle, triangle), non-figurative (with a few exceptions, e.g. airport) and can be applied in many ways. The linear ones have been the object of methodical research (simple, double, dotted line). Those expressing dynamics can be matched with a multitude of arrows or be unrolled into strips. For the flat areas expressing the surfaces, it is possible to play with transparency, frame overlays or color gradients.

This library is all the signs at my disposal. In isolation, none of them has any meaning; they are only categories of graphic forms that will be suitable to illustrate objects or intentions. They will be chosen to signify and translate elements at a given moment. It is only when they are placed on the map and captioned in a particular context that they become meaningful.

For example, I like to use the square to symbolize cities: the rigidity of the square and its stability on its base inspire me with the constructed and anchored character of the positioning of a city, whereas in the usual imagery, the city is instead represented on a map by a round point. In prospective maps, the square also marks the voluntary act in the making. In the case of differentiated scenarios, it is necessary to maintain consistency from one scenario to the next, so the city squares must be used in an analogical manner so as not to hinder the reader's understanding.

9.3.3. *Translating, visualizing and expressing intentions*

Often, cartographers work alone: many of the choices they have to make are the result of their own thinking, informed by their own skills. However, being part of the DATAR teams and having to contribute to the formulation of regional planning policies has made the process iterative. This did not only consist of adjusting the representation as discussions went on. I participated in the actual development of the scenarios, some of the intentions that guided them being sometimes contradictory. Their cartographic translation often made it possible to stage these contradictions and clarify the objectives and terms of the reflection.

But the back-and-forth nature of the iterative process requires constant adjustments to the representation, with the risk of damaging the graphic coherence of a proposal. Under these conditions, I sometimes had to impose my views and make my partners understand that what was important was the legibility of the map and the correct translation of the major ideas.

Indeed, one of my goals has always been to make my maps immediately understandable. As in any map, there are always several levels of interpretation in a prospective map; but here the essential political messages must be perceived at first glance. The legends are then used to clarify the intentions. In the case of multiple scenarios for the same territory, as in the case of the White Paper, each of the maps includes a few invariants of the representation: the boundaries, the cities, the transport routes, the forests, etc., remain the same from one scenario to the next. What changes from one scenario to another is the importance that will be given to each of these elements, one in relation to the other, according to the development choices expressed. Each scenario map must be distinguished from the other two in order to illustrate the diversity of political choices likely to lead to contrasting territorial configurations. Each one must provide at least two levels of reading: a general meaning – the essential message that will differentiate the overall configurations – and more detailed elements of analysis of the components of differentiation.

9.4. The first maps of the great national debate on regional planning

In the early 1990s, DATAR also contributed to the National Debate on Spatial Planning, designed by the French government as an exercise in collective reflection on the future of the French territory at the dawn of the 21st century. In 1993, a DATAR team invited me to work with them on the many forms of representation that should accompany and clarify this debate.

An initial objective was set for me: to depict the urban and territorial organization in the France of the future that is not exaggerated and could be validated at all institutional levels. To do this, our team had a large number of studies at its disposal, including the reports of the working groups of the seven regional projects, and that of the Paris Basin, the statistical work of the *Institut National de la Statistique et des Études Économiques* (INSEE), and the work of the academic Damette (1994) on the links between cities, in particular the material links (exchanges of goods and passengers, rail and road) and immaterial links (banking and telephone flows). The DATAR project managers were also questioned about urban complementarities in their specific field: economy, health, education, etc. From this material, the aim was to define the outlines of urban systems capable of optimizing the functioning of the whole territory. My mission was mainly to find new graphic forms of representation to highlight the idea that large, medium and small cities irrigate the whole national space by playing on their complementarities.

The example of the representation work carried out for the Paris Basin appeared to the team as a model. However, it was decided to dispense with the scenario method and to retain only the desirable future of the connected metropolitan network. The introductory document to the debate, published in 1993, includes many analysis and synthesis maps. But it is mainly the maps I produced on the "long-term outline" which have fed the debate (see Figure 9.3).

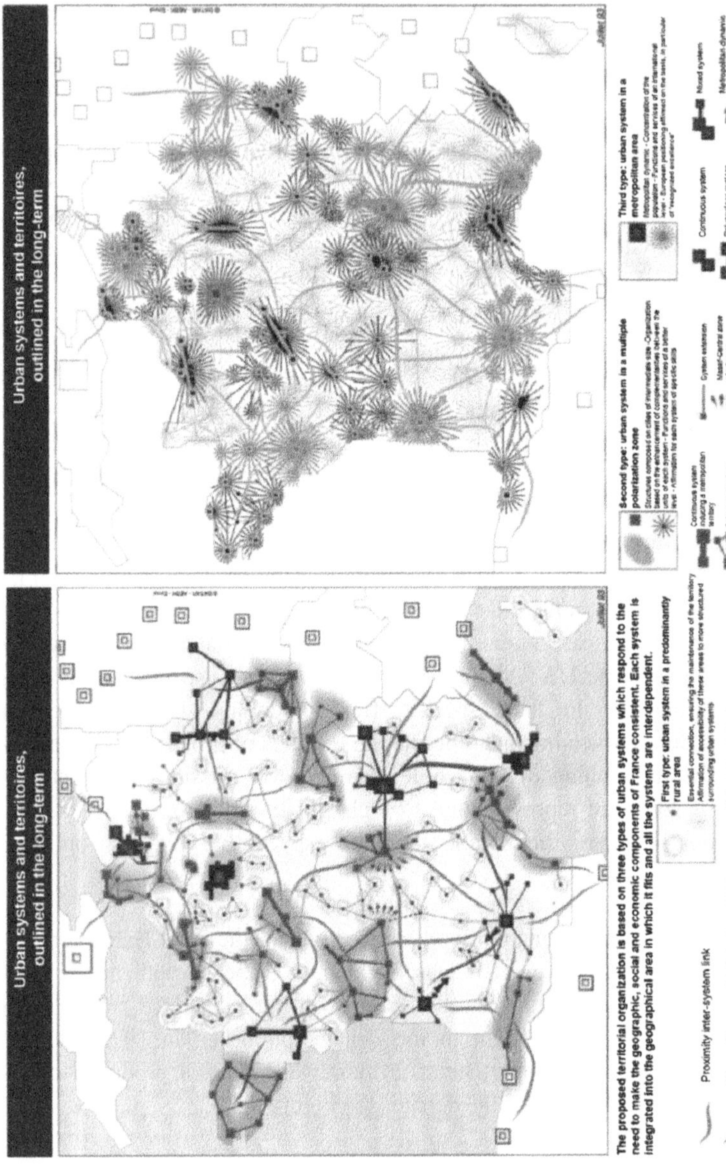

Figure 9.3 *Urban systems and territories, 1993. Long-term outline (source: DATAR (1993, p. 54))*

These are in fact two different maps, sharing a single legend. The first highlights the urban systems (in terms of what was done for the Paris Basin) likely to present an alternative to the polarization deemed excessive by the Paris capital on the scale of the entire national territory, but also to withstand European competition. Their design is based on statistical, economic and transport flow analyses illustrating the links between the main cities. The second takes into account the interweaving and continuous nature of the territories that encompass these urban systems. The latter are classified into three types: urban system in a predominantly rural area, urban system in an area of multiple polarization and urban system in a metropolitan area. Each of the three is precisely defined in the legend. Whole systems cover the entire territory.

The construction of these maps was collective once again. As a geographer and cartographer, I participated in all the meetings to define the systems before working on their representation. There were many questions: Which cities or groups of cities identified in each of the seven major regional projects can claim to be European-level metropolises? What kinds of links exist between them? What complementarities can be developed? For the multiple polarization systems, a threshold of 500,000 inhabitants was set so that together the cities making up the system could play a unifying role in the territory. Finally, which were the motivated small towns in the rural environment and in what configuration could they, together, energize the interstitial territory of the other systems? All these questions were asked and there is no graphic link between towns on the first map that was not discussed in this initial version.

I drew several sketches by hand before finalizing the chosen one in DTP. The size of the cities corresponds to their population class. The size and thickness of the bars connecting the cities are a function of the various complementarities that were taken into account. The graphic representation is also schematic here. The same applies to the second map, the objective of which was to show the limits of territorial influence of the cities, regardless of their size. This map, known as the "urchin" map, was very quickly denounced because it was misunderstood. It was not so much the shape of the urchins, which at the time was a completely new method of representation facilitated by digital cartography, that caused debate but the fact that the intermediate level of the multiple polarization systems stood out more than the metropolitan systems because it was darker in color! Cartographic heresy? No, as I have explained (Bailly 1993), attention was deliberately drawn to these systems because they were the newest and the ones on which the state had to convince the cities to play complementarity. At the time, it was not easy to get Rouen, Caen and Le Havre, Reims and Châlons-en-Champagne or Poitiers, Limoges, La Rochelle and

Angoulême to work together. My role as a cartographer was to specifically point out this issue.

Launched under the aegis of the Prime Minister and led by the Minister of the Interior, the national debate took place with the help of a document containing these maps (DATAR 1993). The prefectures were mobilized. The local authorities actively participated and debated. Numerous modifications of the content, notably on the links or the configuration of urban systems, were requested. The maps had to be modified according to the wishes of the politicians coming from the field. The basic data were handed over and the cartographer complied with the requests. The final maps had to be, first and foremost, the product of the debate, a continuity of their iterative construction.

9.5. France in 2015

At the beginning of 1994, DATAR, at the request of the government, went back to the drawing board and proposed a new step document: "France 2015". This time, the aim was to provide French policy with a territorial vision capable of framing the state's sectoral policies and the policies of regional and local authorities (DATAR 1994a). The two controversial maps in the introductory document mentioned above were replaced by three maps with a new graphic design, each accompanied by a European label that sustained their foothold. The aim was to fill a gap in the previous exercises and to remind the reader that the organization of the territory of neighboring countries had an influence on that of the French territory.

The map entitled "Organisation of the territories in 2015" (see Figure 9.4, left-hand map) shows the seven major territorial projects selected, differentiated by dominant colors. I have proposed a system of alternating bands of varying density to mark the influence of the major urban centers responsible for influencing these territories. For the areas on the fringes of these groups that are influenced by several hubs, the alternating bands are finer and they intermix. This map was difficult to develop and took a long time. Iterative cartography requires a lot of patience. Each vertical line is composed of sections whose thickness, length and color were the result of laborious discussions. They had to be modified one by one and each modification led to a cascade of others.

The second map, called "Architecture of Cities 2015" (see Figure 9.4, right-hand map), symbolized the relationships between cities, of all kinds, tangible and intangible. It took up the idea of the urban framework, already designed in 1993, but the links were largely reworked and the original urban systems connected to each other.

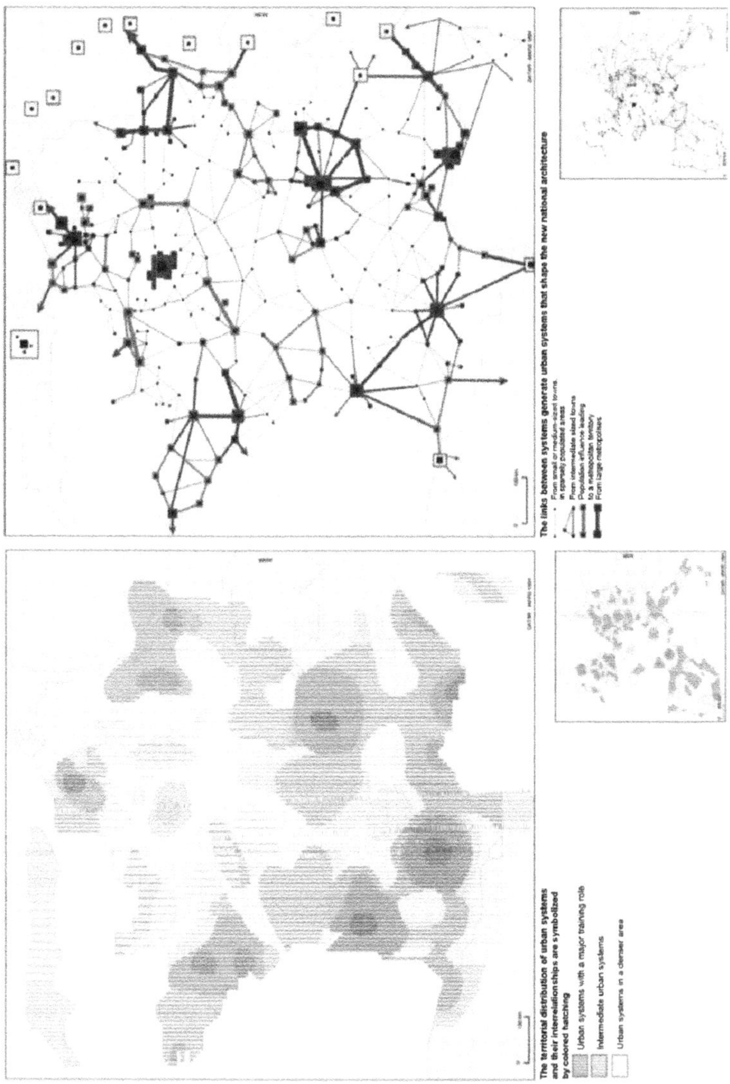

Figure 9.4. *A new face for France in 2015, territorial organization and city architecture (source: DATAR (1994a, pp. 18–19))*

The introductory document (DATAR 1993), a proposal from the state, was intended to be directive. Paradoxically, politicians and public opinion have simplified the discourse and made it more descriptive (DATAR 1994b). This work was the subject of a charter but the changes in the French Presidency and in the political majority in 1995 – both of which shifted to the right – did not allow the process to continue toward a law of programmatic orientation.

9.6. France in 2020

In 2000, Jean Louis Guigou was formally appointed Delegate at the head of DATAR. I had worked with him in 1995, as Director of European Affairs, when he was following the ESDP (European Spatial Development Perspective), a reflection on the future of the territory of the European Union, for which I had produced the trend scenarios maps of the ESDP (Spatial Development Committee 1995). Guigou was a fervent supporter of foresight; he liked maps and he intended to re-launch the debate on the future of the French territory, the role of cities and their integration into an expanding Europe.

In 2000, the context was no longer the same. Following the decentralization laws of the previous decades, the French State had become one of several actors in the country's spatial planning policies; this is why this exercise for a "France in 2020" (DATAR 2000) had become multi-scalar, and DATAR's visions had become more drawings than designs. However, the Delegate had a clear and directive vision of the plans he wanted to popularize. DATAR and the cartographer resumed their collaboration.

Guigou's idea was to return to the scenarios method and to retain four of them: three scenarios playing a deterrent role, followed by a fourth, which appeared to be the only viable one. For the cartographer that I was, and despite all the sympathy I had for this new Delegate who was entrusting me with the task of representing these scenarios, the order seemed a little tough to me. We were in pure fiction. It was a question of translating four ideas without much connection to the geographical reality and the available data.

In order to respond to this new request, I used the graphic outlines of France previously used by DATAR, increased by the border countries and cities. I used previous maps to locate the cities. These were the only invariants of the four scenarios. Note in passing that the representation of the cities is no longer square as in the previous exercises but takes the traditional shape of a circle or sphere.

9.6.1. *Three "deterrent" scenarios*

The first three scenarios were designed in a caricatured manner, intended to point out threats to the balance of the French territory (see Figure 9.5). Scenario 1, the so-called "renewed centralism" scenario, simulated a return to the dominant role of central government and Paris. I used the symbolism of the "Legrand Star", a term that refers to the structure of the French railway network designed in the 1850s by the engineer Legrand, in response to the Second Empire government's request to use the network to reinforce the centralization of the country in Paris. This choice was the first scenario to be rejected, illustrating a step backwards in the way of thinking about the territory of a country that had since become largely decentralized. Scenario 2, known as the "fragmented archipelago", gave the image of a fractured France: the most dynamic centers were competing with one another and looking to their foreign neighbors. Scenario 3, known as the "differentiated local" scenario, presented a France where local authorities, at all levels, work in a dispersed manner, leaving the less dynamic ones on the sidelines. This map was the most difficult to construct and the only one that required a legend to be understood. The graphic approach adopted also illustrated the repulsive function of the corresponding scenario.

9.6.2. *The "connected polycentrism" scenario*

Finally, scenario 4 (see Figure 9.6), the one promoted by DATAR, known as "connected polycentrism", was a graphic reproduction of the 1994 urban architecture. But make no mistake: the links drawn between the main cities were fictitious; they were not based on statistical or scientific studies. Only the idea of polycentrism that guided the thinking in France and in Europe at that time prevailed (Debarbieux and Vanier 2002). I had to modify this map many times because it was necessary to combine this representation of the urban network with that of the six large territorial basins, reminiscent of the main hydrographic basins of the country, to which Guigou was attached. As these large basins were not unanimously accepted, it was necessary to reduce their visibility; initially differentiated in large areas of bright and contrasting colors, they were finally represented using a single blue color as a background. The final version, which was not the most popular, except on a stamp commemorating the 50th anniversary of DATAR, even made the colored background disappear, leaving only the canvas of cities and links visible.

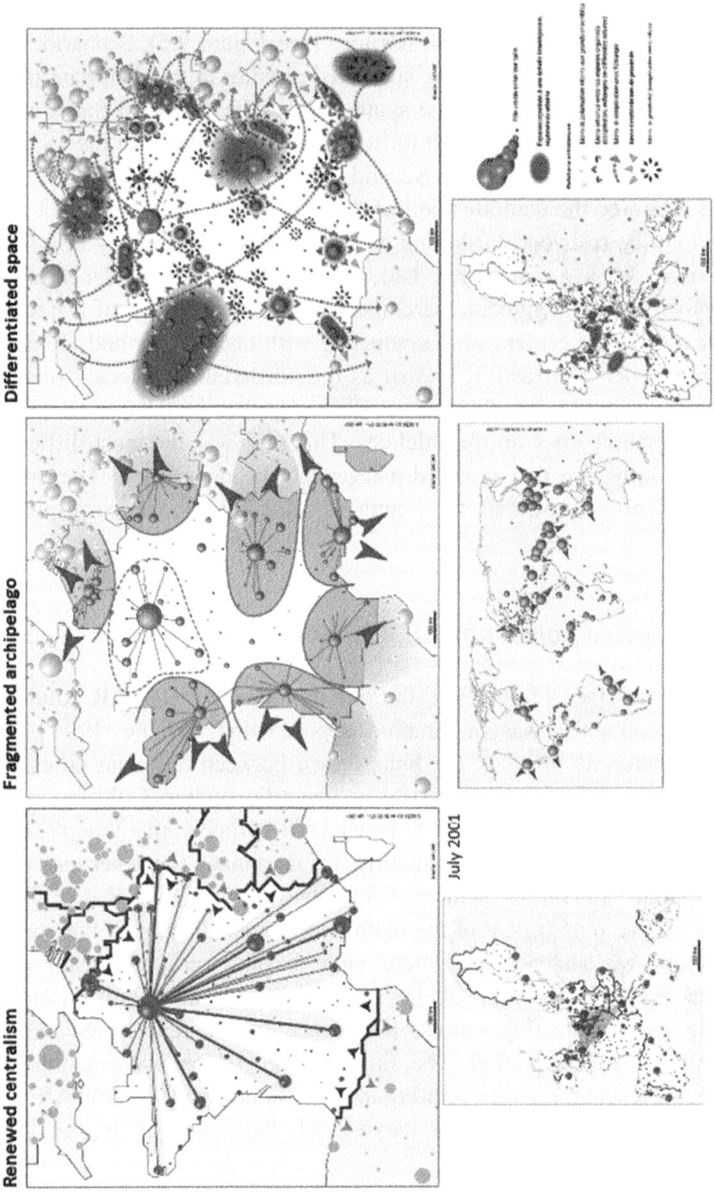

Figure 9.5. *Three scenarios for France in 2020 (source: DATAR (2000, pp. 56, 58, 60))*

Scenario 4

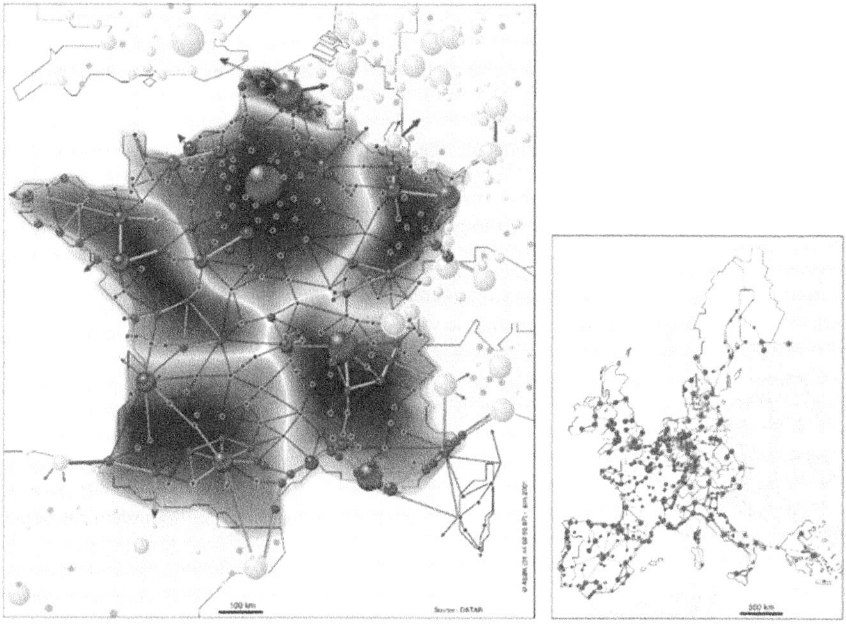

Figure 9.6. *Meshed polycentrism for France in 2020 (source: DATAR (2000))*

9.7. Lessons from a wealth of experience

Prospective cartography in spatial planning is not expert cartography contributing to an inventory or diagnosis, nor even project cartography. It gives a visual form to the objectives that guide policies, facilitating their understanding, animating debates and leading to their adoption. It is therefore eminently political in itself. As such, it cannot be said to be a data-driven map. In this, I agree with Claire Bretécher's map-maker.

9.8. References

Bailly, A. (1993). Aménagement du territoire et cartographie : une même révolution. *Lettre de la DATAR*, 146.

Bertin, J. (1967). *La sémiologie graphique, les diagrammes, les réseaux, les cartes.* Gauthier-Villars, Paris/The Hague.

Brunet, R. (1986). La carte-modèle et les chorèmes. *Mappemonde*, 86(4), 2–6.

Centre de Création Industrielle (CCI) (1980). *Cartes et figures de la terre*. Centre Georges Pompidou, Paris.

Cercle pour l'aménagement du territoire (ed.) (2016). La DATAR, 50 ans au service du territoire. French document, Paris.

Conférence permanente des Présidents de Région/DATAR (ed.) (1994). Charte du bassin parisien. DATAR, Paris.

Comité de Développement Spatial (1999). *SDEC. Schéma de développement de l'espace communautaire, vers un développement spatial équilibré et durable du territoire de l'union européenne*. Commission Européenne, Luxembourg.

Damette, F. (1994). La France en villes. French document, DATAR, Paris.

DATAR (ed.) (1971). Une image de la France en l'an 2000. Scénario de l'inacceptable. French document, Paris.

DATAR (ed.) (1992). Livre blanc du bassin parisien. French document, Paris.

DATAR (ed.) (1993). Débat national pour l'aménagement du territoire, document introductif. French document, Paris.

DATAR (ed.) (1994a). Débat national pour l'aménagement du territoire, document d'étape, la France de 2015. French document, Paris.

DATAR (ed.) (1994b). Débat national pour l'aménagement du territoire, synthèse du document d'étape. French document, Paris.

DATAR (ed.) (2000). Aménager la France de 2020. Mettre les territoires en mouvement. French document, Paris.

Debarbieux, B. and Vanier, M. (2002). *Ces territorialités qui se dessinent*. Éditions de l'Aube/DATAR, La Tour d'Aigues.

Duport, J.-P. (1993). Une nouvelle architecture pour la France. *Lettre de la DATAR*, 144.

Guigou, J.-L. (1993). *France 2015. Recomposition du territoire national*. DATAR/Éditions de l'Aube, Marseille.

Guigou, J.-L. (1995). *Une ambition pour le territoire*. DATAR/Éditions de l'Aube, Marseille.

Guigou, J.-L. (2008). Une égo-histoire de la DATAR. In *Colloque de Cerisy – Aménagement du territoire, changement de temps, changement d'espace*, ALLEMAND, S., FRÉMONT, A., HEURGON, E. (eds). Presses universitaires de Caen.

10

Making Maps to Fight Unjust Migration Policies

Olivier CLOCHARD
CNRS, Migrinter, University of Poitiers, France

10.1. Introduction

Migreurop is a Euro-African network of international associations and individual members, created in 2002, which focuses on the various forms of confinement of migrants as well as the deployment of migration controls at the borders of the European Union (EU) and in neighboring countries. Through press releases, reports and books, the network aims to gather information on migration situations that are difficult to understand. Starting from a reflective approach to the cartographic practices that have been deployed within the network, this chapter aims to describe the way in which maps are produced and to examine the ways in which illustrations contribute to the criticism of European migration policies, while at the same time providing knowledge on the effects of these policies on the migrants' journeys. The aim is to understand how maps – some of which sometimes tend to break with cartographic conventions – and their evolution can help us better understand migration dynamics. In the first part, we seek to answer the following questions: How are these visual regimes constructed? At what level does is politics involved in cartographic production and dissemination? In the second part, we present the main dimensions of the maps produced by Migreurop, which seek to capture politics and policy, whether that is to provide information and support claims, to seek new modes of representation, to give importance to the words used or to collectively design sensitive, informative and artistic creations.

The Politics of Mapping,
coordinated by Bernard DEBARBIEUX and Irène HIRT.

10.2. Actors and objectives of cartographic productions

Migreurop's actors produce maps from a variety of sources of geographical information and from various collaborations between cartographers and non-cartographers, all processes that support the network's struggles against European migration policies that push many people into inhumane and degrading situations.

10.2.1. *From respect to emancipation of semiological conventions*

Network members who are not trained in cartography, but who have a good knowledge of migration issues, sometimes use spatial language because of their proximity to cartographers who have rallied together to improve living conditions for migrants; this language, in some cases, has been the source of various (carto)graphic inspirations or drawings, without the substance of the map being lost.

For example, the map "EU-Africa: a patchwork of cooperation frameworks to 'manage' migration" (see Figure 10.1) is the result of a patient mapping of the different types of migrant readmission agreements, an activity carried out since the mid-2000s within the network. These agreements, concluded in a bilateral, intergovernmental or community framework under the leadership of EU countries, aim to have external states readmit their nationals, or even people who have transited through their territory, who are stopped in an irregular situation on EU soil (Migreurop 2012). This work has also revealed that it is very complicated to have access to the text of a bilateral readmission agreement signed between a Member State and a third country, such as those concluded between Italy and Libya (Charles 2007). Moreover, as some readmission clauses are integrated into political, commercial or economic agreements, the inventory cannot be exhaustive. In the end, this map only partially captures the extent of the spatial scope of the policy on readmission agreements, which enable EU-Africa cooperation to be based on increasingly stated objectives, namely "to stem the [imagined] influx of irregular migrants" and "to increase the rate of return to countries of origin and transit" (European Commission 2006).

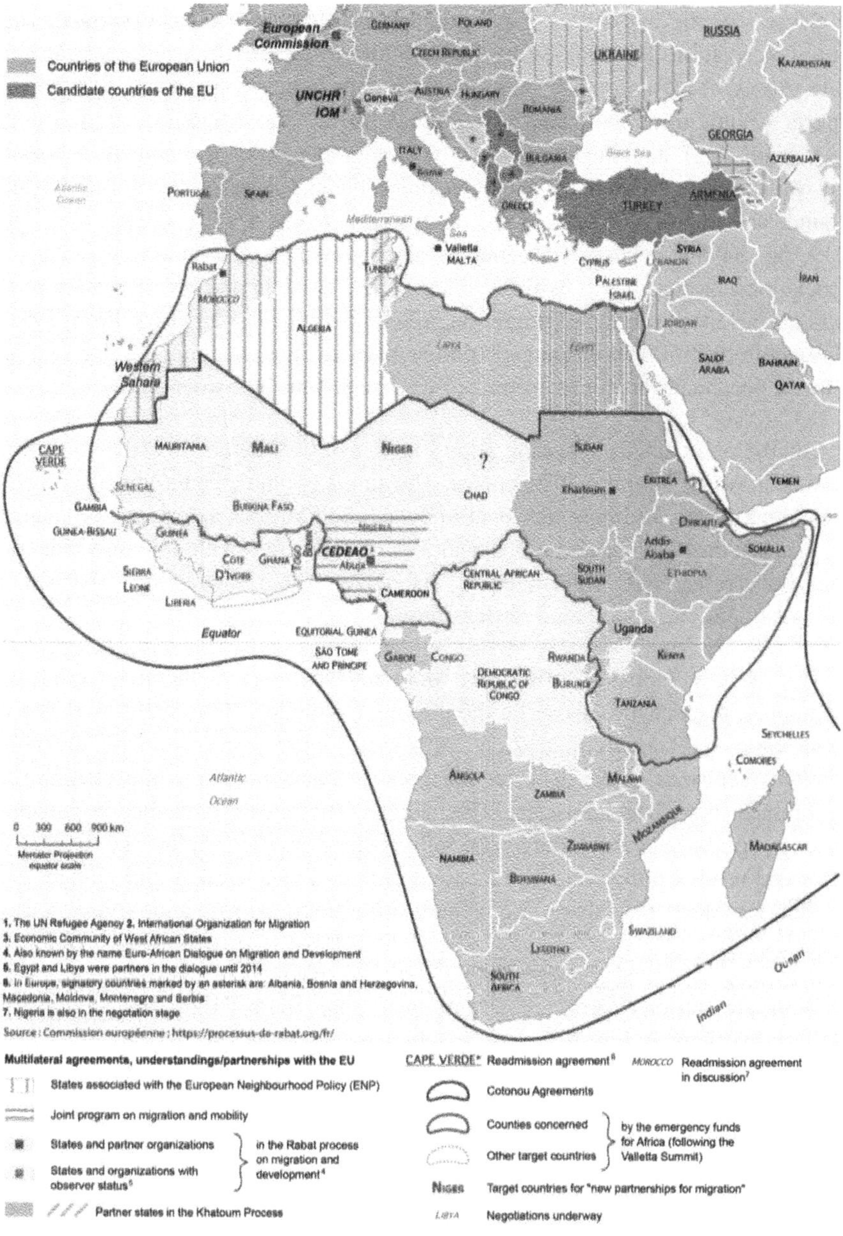

Figure 10.1. *EU-Africa: a patchwork of cooperation frameworks to "manage" migration (source: map by O. Clochard, C. Charles and P. Chappart for Migreurop in Migreurop (2017, p. 75))*

Several of the authors of these productions also use their commitments as a source of creation. By thinking of the map in a different way from a single *Excel* table, this type of approach leads to a strengthening of the criticism of certain aspects of migration policies. The map entitled "The Migratory Red Mound" (see Figure 10.2), imagined and drawn by cartographer Nicolas Lambert, echoes the anti-war song composed in 1923 by Gaston Montéhus, in "reference to the Bapaume Mound, one of the bloodiest places of the First World War" (Lambert and Zanin 2019, pp. 94–95). Thus, it is a matter of conveying a conviction shared by the associations, according to which violence against migrants characterizes at the highest level the evolution of migration policies over the last 30 years. From 1993 to 2019, more than 50,000 migrants lost their lives at European borders, due to a lack of travel documents and/or as victims of warlike actions by border police.

Finally, participatory mapping allows us to ask field actors or exiles to describe the way in which migration takes place. The aim is to gain a better understanding of lived experiences, and how the local knowledge of various actors can be mapped. We are then deal with spatial representations of collectives with drawings that come from both researchers and migrants. Moreover, these interactions sharpen the analyses of policies and related responsibilities.

This is how Thibaut Duffey and Morgane Dujmovic, a mountain guide and geographer, respectively, describe the migration situation "from Ventimiglia to the Roya Valley" (Migreurop 2017, p. 109) (see Figure 10.3), showing the police resources deployed since 2011 on the French–Italian border, while recalling the natural elements of the landscape. The hand-drawn map, accompanied by a caption with a long text and three drawings, tells readers how legislations go against the rights and needs of migrants. The location of the various hamlets and villages is an implicit reminder that there are people who help exiles facing these various obstacles.

With the various forms of commitment that characterize the network, the members participating in these cartographic constructions aim to give more means of understanding to migrants and the citizens who support them. Consequently, with these maps in hand, all the people in a way become mediators of cartographic sciences without necessarily being experts. But to achieve this, it is necessary to have structuring elements to guide their actions.

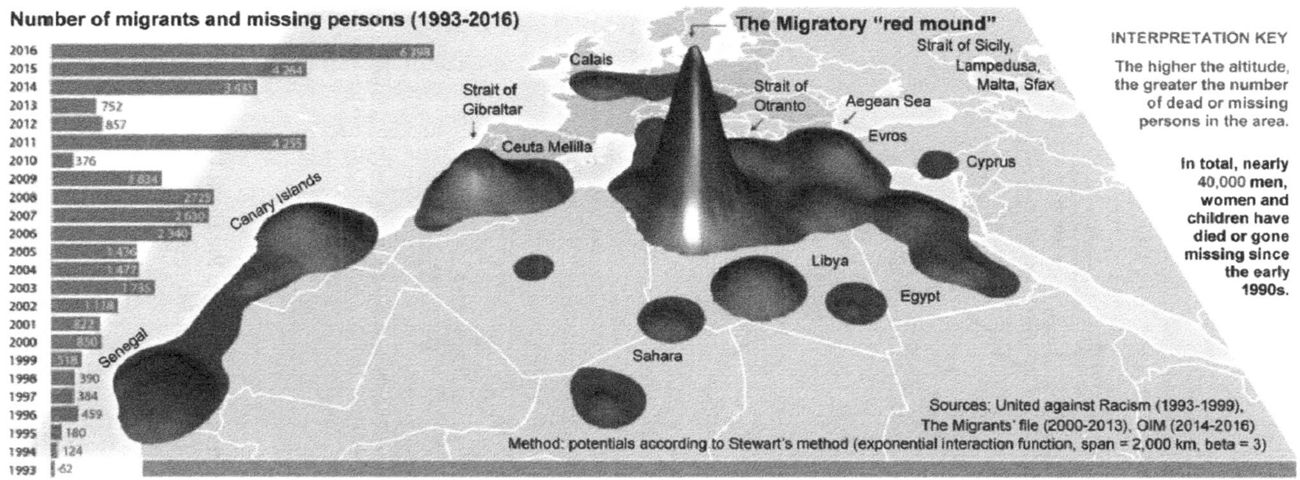

Figure 10.2. *"The Migratory Red Mound" (source: map by N. Lambert for Migreurop in Migreurop (2017, p. 128)*

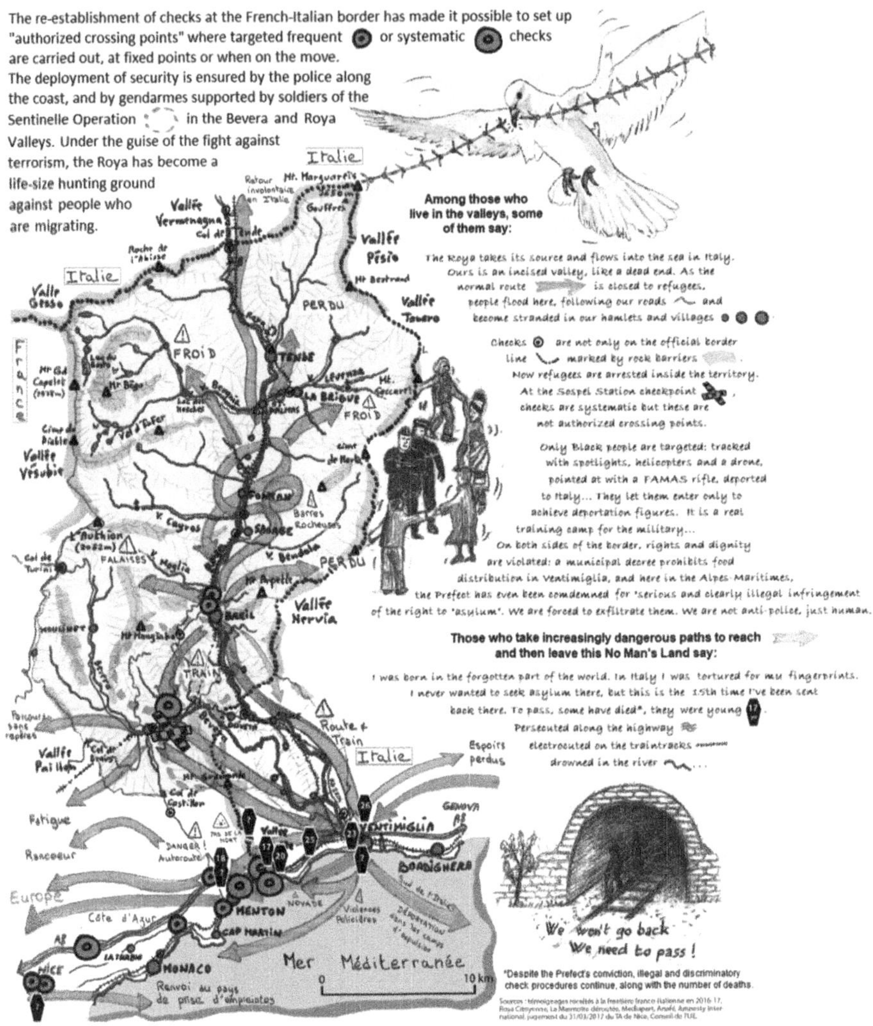

Figure 10.3. *From Ventimiglia to the Roya Valley (source: map by T. Duffey and M. Dujmovic for Migreurop in Migreurop (2017, p. 109))*

10.2.2. *The two network compasses*

Even if we know that mapping is only one of a variety of means to answer the questions that many people ask, maps have the advantage of being able to make "elements that could be presented in a scattered way in a verbal presentation" coexist (Levy *et al.* 2004, p. 3). The documents produced are thus elaborated collectively to help associations, activists, teachers, etc., to recognize themselves in the ideas defended by the network, and to be the basis of debates, much more than to impose thought patterns on them. It is also a question of questioning situations and principles that the media take little interest in, and of re-examining our ways of thinking and doing, so that people can act with full knowledge of the facts. In this way, we strive to "modify the collective representations that are transmitted by institutions" (Noiriel 2010, p. 219).

Based on the observations of field actors (associations, activists and researchers) who understand the difficulties that many migrants face, especially when they are detained, it became clear from the very beginnings of the network that a political position was needed to combine action and research. Thus, a first text accompanied the map of the main places of administrative detention of migrants in Europe and beyond (see Figure 10.4), introducing the word "camp" into the public debate (see below). Then a European appeal, issued at the initiative of Migreurop in 2004, opposed the creation of camps at Europe's borders, envisaged at the 2003 Thessaloniki European summit. In opposition to the "Return" directive[1], Migreurop (2010) also asked "the governments of the EU Member States and of the countries located at its borders to stop using detention for the purpose of migration control, and called on public opinion to reject all logics of confinement of migrants". The administrative detention of migrants, by its very nature, leads to the violation of fundamental rights, and often escapes any effective jurisdictional control, while individual freedoms are at stake. From 2009 to 2014, campaigns to visit detention centers[2] were added to this political position in order to gather information on these facilities.

1 European Directive of December 16, 2008 on common standards and procedures in Member States for returning third-country nationals without the legal right to remain.

2 Most of these visits were carried out with national or European parliamentarians who are normally guaranteed access to these establishments.

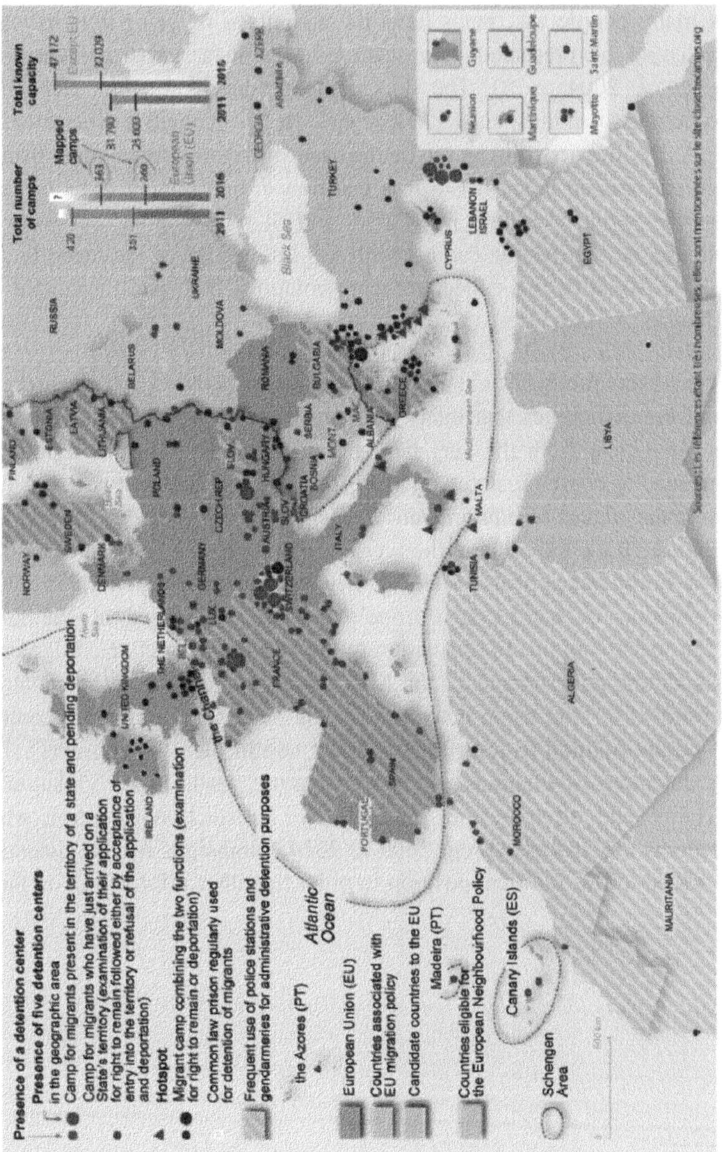

Figure 10.4. *The European Union's camps (source: map of Migreurop members for Migreurop in Migreurop (2017, p. 43))*

By recalling, on the one hand, that migrant camps are places of confinement for people to whom the authorities deny the right to settle in their country and, on the other hand, that the Universal Declaration of Human Rights (UDHR)[3] advocates the right to leave, the network concludes that the right mentioned in the UDHR is strongly undermined if people do not have the right to arrive, and therefore to settle where they wish. The question of the free movement of people thus arises at the end of this reflection and the network pronounced itself in favor of it in 2013. This demand "in the name of the principle of equality, of the freedom to come and go and to settle for all, constitutes the indispensable corollary of the defence of migrants' rights [economic, social, etc.]" (Migreurop 2013).

10.3. Mapping experiments undertaken

Since the creation of Migreurop, the map has been one of the network's structuring elements, in the sense that it has made it possible to disseminate knowledge and claims about the measures relating to the control of migratory flows in a social environment (associations or groups supporting exiles, etc.) that made little use of the spatial medium of the map until the early 2000s. Based on various publications and the *Moving Beyond Borders* exhibition (see section 10.3.4), the aim is to analyze the way in which European migration policies are implemented and their consequences, as well as to examine the responsibilities of governments. By giving free rein to new modes of graphic, visual and multimedia representation of migration policies and their effects, certain maps have adopted or participated in the dissemination of new forms of vocabulary that are sometimes seen as provocative by governments and organizations less inclined to criticize migration policies. However, the scenarios that the words foreshadowed have been all too true in these new media for disseminating knowledge. Finally, the network members involved in these cartographic constructions have also sought to inject creativity, both in content and form, with the idea that these issues would become better known and understood by the general public.

10.3.1. *Carrying knowledge and claims*

The "power of maps", an expression often used since the work of Brian Harley, "shows that the 'simple' act of naming or locating an element on a map often has a

3 "Everyone has the right to freedom of movement and residence within the borders of each state. Everyone has the right to leave any country, including his own, and to return to his country" (UDHR, Article 13).

political impact" (Gould and Bailly 1995, p. 21). This is precisely what Migreurop undertook to do from the summer of 2003 onwards by mapping the places where migrants are detained in all EU Member States and neighboring countries, in order to show the growing importance of these devices in European policies (see Figure 10.4). The aim was also to denounce the consequences of these camps, which are detrimental to the rights of migrants (refusal by the authorities to register an asylum application, failure to respect people's private and family life, etc.), and even to the rights of children, with international texts recommending that states have exceptional recourse to the administrative detention of foreign minors (Clochard 2018).

From an intuitive and empirical approach, a first database composed of eleven criteria[4] was established with information from official statistics, parliamentary reports or those made by voluntary associations and observations made by activists, associations and researchers for the network. The map produced from these data was a reminder of a reality that many European governments do not want to show; the segregation of migrants in degrading and inhumane places. Nevertheless, this first map produced by Migreurop could be seen as a truncated and biased geographical tool of reality (Rekacewicz 2006) because, in the absence of sufficient data, it did not cover all European countries and neighboring states. However, it has become a valuable tool for knowledge about the places where migrants are housed. In its sixth version published in 2016, the map of camps in the European Union shows that while they decreased in number from 420 to 360 between 2011 and 2016, this did not reflect a "decrease in the detention of migrants but a rationalization process linked to state objectives of minimizing expenditure and the increasing privatization of this sector" (Migreurop 2017, p. 42).

4 Geographical coordinates of the camp; name; capacity; annual number of people detained (men, women and children); whether it is a migrant camp for the purpose of examining their application for admission to the territory of a State, a migrant camp present within the territory of a State and awaiting deportation, or even a camp combining both functions; date of opening; if applicable, date(s) of closure; maximum length of detention; the different actors running the camp (State, private group, NGO, etc.); presence or not of a legal association; presence or not of a humanitarian association.

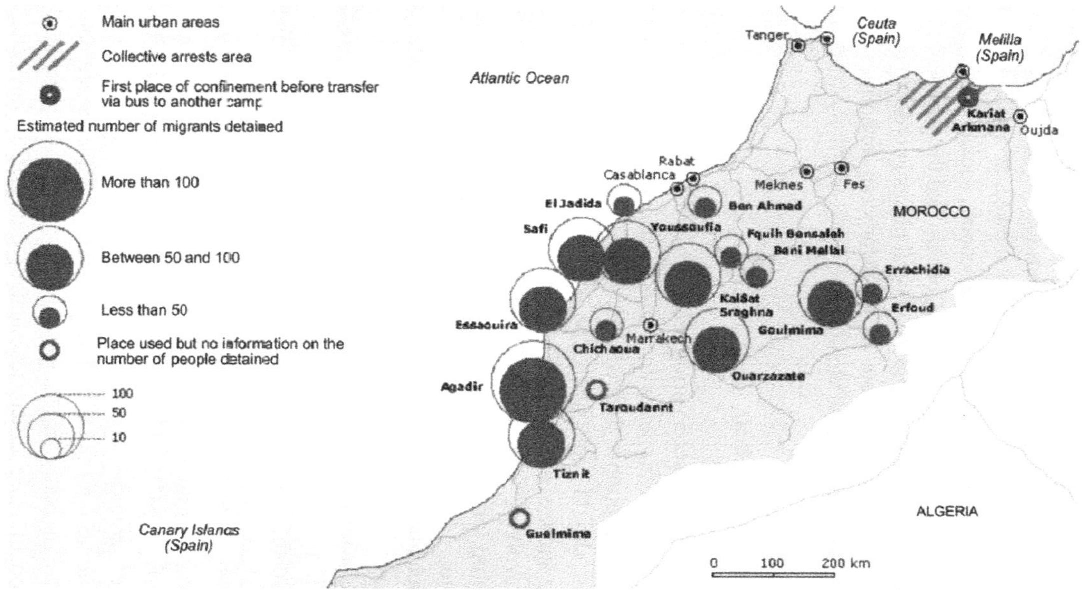

1. Kariat Arkmane is used to identify the people arrested (photographs, fingerprints, marital status, etc)
2. Statistics relating to migrants detained in each camp have been grouped into 4 groups. Faced with the difficulty of obtaining precise figures, each time we have put a dotted red circle to remind us that the number of migrants imprisoned has often been higher
3. People have generally been released following 3 weeks to a month of arbitary detention

Source: joint CCSM-GADEM briefing note on the arbitrary displacement and detetion of migrants in Morocco following the roundups of 10 February 2015. Map designed and produced by Elsa Tyszler and Olivier Clochard, 2015

Figure 10.5. *Places of arbitrary detention in Morocco (February 2015) (source: map by O. Clochard and E. Tyszler for the Close The Camps project, https://closethecamps.org/2015/03/20/rafles-et-detention-arbitraire-au-maroc-fevrier-2015/)*

Another example, on the margins of the EU, is the arbitrary detention of migrants in Morocco[5], in the sense that the authorities have not issued a decree legalizing the detention of migrants in an irregular situation. It is difficult to carry out a census of these places. Nevertheless, the map devised in 2015 by sociologist and Migreurop member Elsa Tyszler is symptomatic of the power of maps (see Figure 10.5). For the first time, based on interviews and field research, a map showed, at a minimum, the number of migrants detained in 18 cities in Morocco (in schools, social centers, sports complexes, etc.). It not only had an impact in the media, but also led the president of the administrative court of Casablanca to send a bailiff to several places in order to open an investigation following the roundup carried out by the Moroccan authorities in the camps established near the Spanish enclave of Melilla.

10.3.2. *New modes of representation*

Since 2009, the network has produced various editions of the *Atlas des migrants en Europe* (Atlas of Migrants in Europe) with a group of cartographers who are considering cartographic innovations. Indeed, the latter considers that these cartographic constructions should not solely be the result of "processes of transformation, comprising successive stages, placed in a strict order, integrated into a logical reasoning and involving [scientific] choices" (Cauvin *et al.* 2010). The maps produced by Migreurop, on the other hand, constitute "a very heterogeneous corpus, all of which, to varying degrees, reflect a critical, experimental, 'indisciplinary' or participatory positioning. By combining the scales of geographical analysis, they make it possible to better grasp and develop the characteristic themes of European migration policies. The (carto)graphic writing chosen here is thus the fruit of an assumed subjectivity, the only way to give a place to those who are reduced to categories, flows or faceless silhouettes of surveillance devices" (Migreurop 2017, pp. 9–10).

With the multiplication of resources, especially large databases (United Nations High Commissioner for Refugees [UNHCR], Eurostat, etc.) and the maps that result from them, people's lived experiences can sometimes become blurred. This is why it is also important to take into account the singular. A mapping of a localized path or situation generally leads to a better embodiment of the migration theme. For example, the representation of exile itineraries such as those of Syrian refugees (Lagarde 2019) illuminates these societal facts in a different way. It is a matter of "taking the time to look deeply into the meaning of the lives of fellow human

---

5 As a reminder, France has also for a long time used places on a discretionary basis: the Arenc center (1963–1981), waiting areas in airports before 1992, and administrative detention facilities (*locaux de rétention administrative* [LRA]) in police stations before 2001.

beings" (Springer *et al.* 2018, p. 41) and of those situations anchored in restricted spaces that are significant of wider phenomena. These are all particular facts that help to identify developments in migration policies.

Migreurop has nevertheless chosen to combine the two approaches, first by mapping the themes in their entirety, while being aware that it is difficult, if not impossible, to summarize the complexity of certain aspects of migration policies today. This is why the network wishes to measure the phenomena, where feasible. Then, it seeks to understand them by putting "the focus on a singular situation, a space, a journey illustrating at a finer scale the way policies are deployed, migrations are experienced" (Migreurop 2017, p. 9). With these different cartographies, it is a question, as with photographs, of repositioning the gaze (Berthomière 2012). Indeed, these media allow us to better shed light on the spatial and political approaches to migrations that have spread across the world of our representations at whatever scale, from small intimate spaces to the territory of the street to that of globalization (Simon 2006).

10.3.3. *The words on the maps are also important*

The words used are very often an important component of the graphic semiology of these maps. They indicate the political orientation advocated by the authors with data that is sometimes "subverted to demonstrate the harmful consequences [of migration policies]" (Bacon *et al.* 2016, p. 205). They also strengthen the links between these illustrations and the accompanying texts. Finally, the network's maps have contributed in their own way to the acceptance of the use of contested notions such as the "war on migrants" (Blanchard and Wender 2007) or the "camp" (Clochard 2020).

Several Migreurop associations were indeed at the origin of the (re)introduction of the notion of the "camp" into the field of social sciences in the early 2000s, in order to describe contemporary places of confinement and to reveal the different relationships with those that preceded them. By using the word "camp", the network also sought to counter the euphemisms often used by the authorities to name these places. However, in 2004, the European commissioner in charge of immigration issues, Antonio Vitorino, was not willing to hear it; he found the term unfit for a description and a good understanding of the arrangements put in place by the EU Member States[6]. Yet several publications show that the arrangements of many detention spaces for migrants are nothing short of camps (Agier 2014).

6 António Manuel de Carvalho Ferreira Vitorino quoted by Alexandrine Bouilhet, "L'Europe financera des 'centres' de réfugiés à l'extérieur de l'Europe", *Le Figaro*, October 2, 2004.

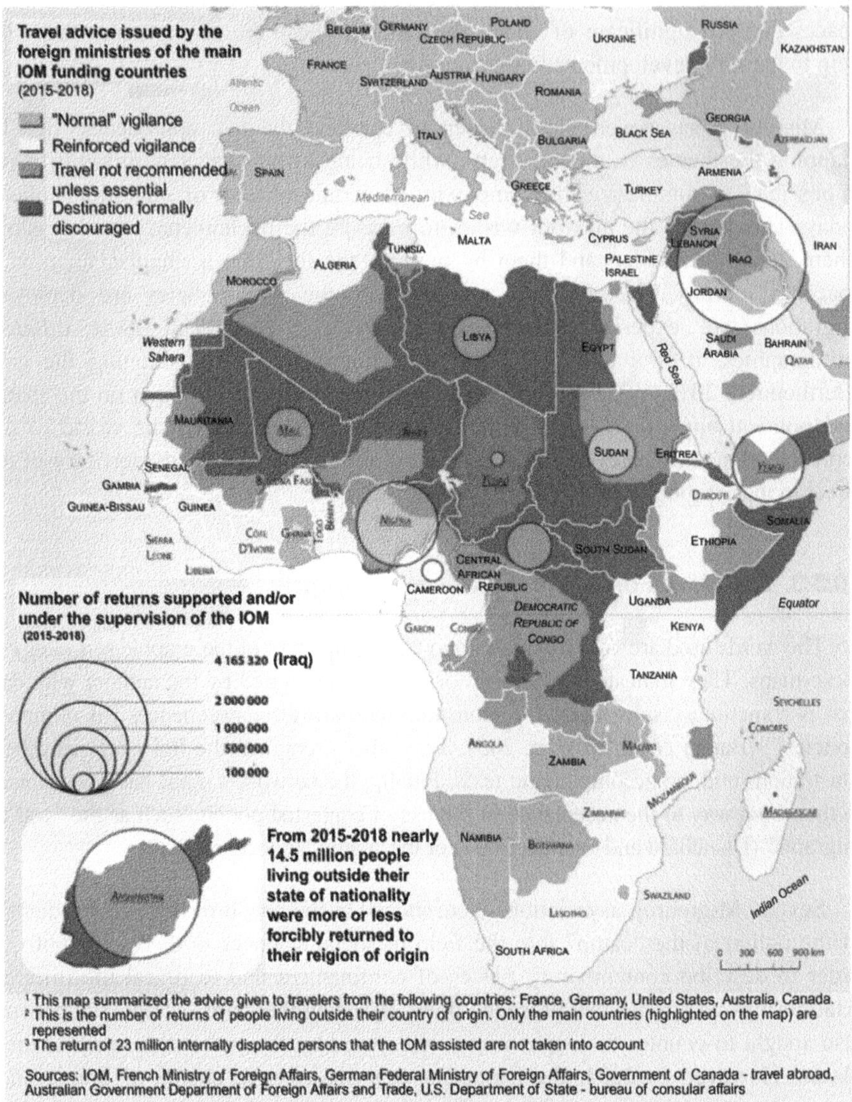

Figure 10.6. *The IOM refers to areas that its sponsor countries consider dangerous (source: map by O. Clochard and K. Rhimou for Migreurop in Migreurop (2019, p. 3))*

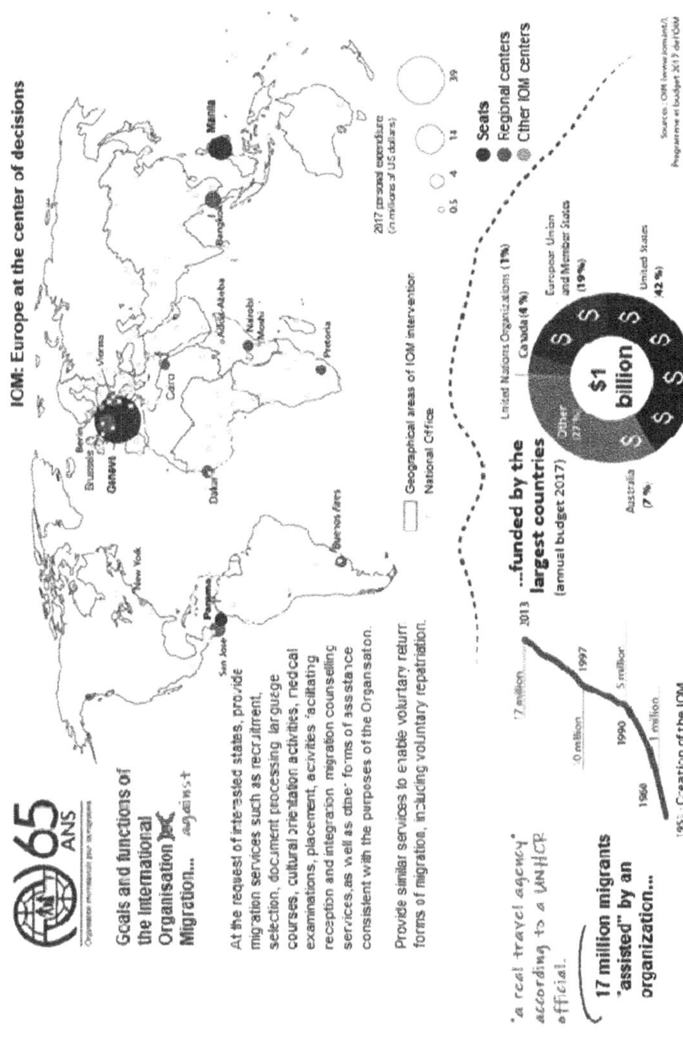

Figure 10.7 The *IOM plays a major role in deportations (source: map by R. Ysebaert for Migreurop in Migreurop (2017, p. 123))*

Words are also important when they are understood by all, or at least most, people. Due to the nature of the network, their work has been extended to other issues, such as the outsourcing of migration controls, women at the EU's external borders, or the role of the Frontex agency and the major international organizations (International Organisation for Migration [IOM] and UNHCR). For example, a map produced by members of Migreurop (2019, p. 3) relating to deportations handled or monitored by the IOM (see Figure 10.6) shows, based on statistics produced by the latter, that it sends people back to countries considered dangerous by some EU Member States including Germany and France. These countries, on the one hand, finance the IOM to ensure deportations to countries in great difficulty (Iraq, Libya, Sudan, etc.), and, on the other hand, they advise travelers not to go there. This "topicality of denunciation" (Boltanski 2007, p. 113) echoes another map by Migreurop (2017, p. 123) on IOM actions (see Figure 10.7) showing that deportations play a major role. The authors of the map have substituted the "for Migration" in the organization's title with "against", quoting a UNHCR representative in Niger who said that the IOM is "a real travel agency".

10.3.4. *The issue of esthetics*

With approaches that are scientific, militant and artistic, the estheticism of the drawings and maps is strengthened without devaluing the contribution of knowledge on migration, or even of the policies implemented in this regard. For example, based on the hypothesis defended by a large body of research that migrants cannot always be assigned to a single institutional and administrative category, the idea was to show the porosity that exists between these classifications: a person may be an asylum seeker before being rejected, and then obtain a "private and family life" residence permit; another may be a refugee and a student at the same time, etc. It is almost impossible to represent these changes in statistics and maps. This is why the map in Migreurop's atlas, "Without work or without papers, Ahmed's dilemma" (Migreurop 2017, p. 19) (see Figure 10.8), shows a chronological sequence of an individual's journey, which respects the way in which the different territories have been crossed while mentioning the "regular"/"irregular" dichotomy of the person's administrative status. Ahmed's drawings consolidate the citizenship dimension of spatial representation. After much back and forth between cartographers, we could say, paraphrasing Jean Cocteau's quote[7], that "the map has become beautiful when the line is living".

7 Jean Cocteau, "Un dessin est beau si la ligne est vivante", https://www.dailymotion.com/video/xx4re4.

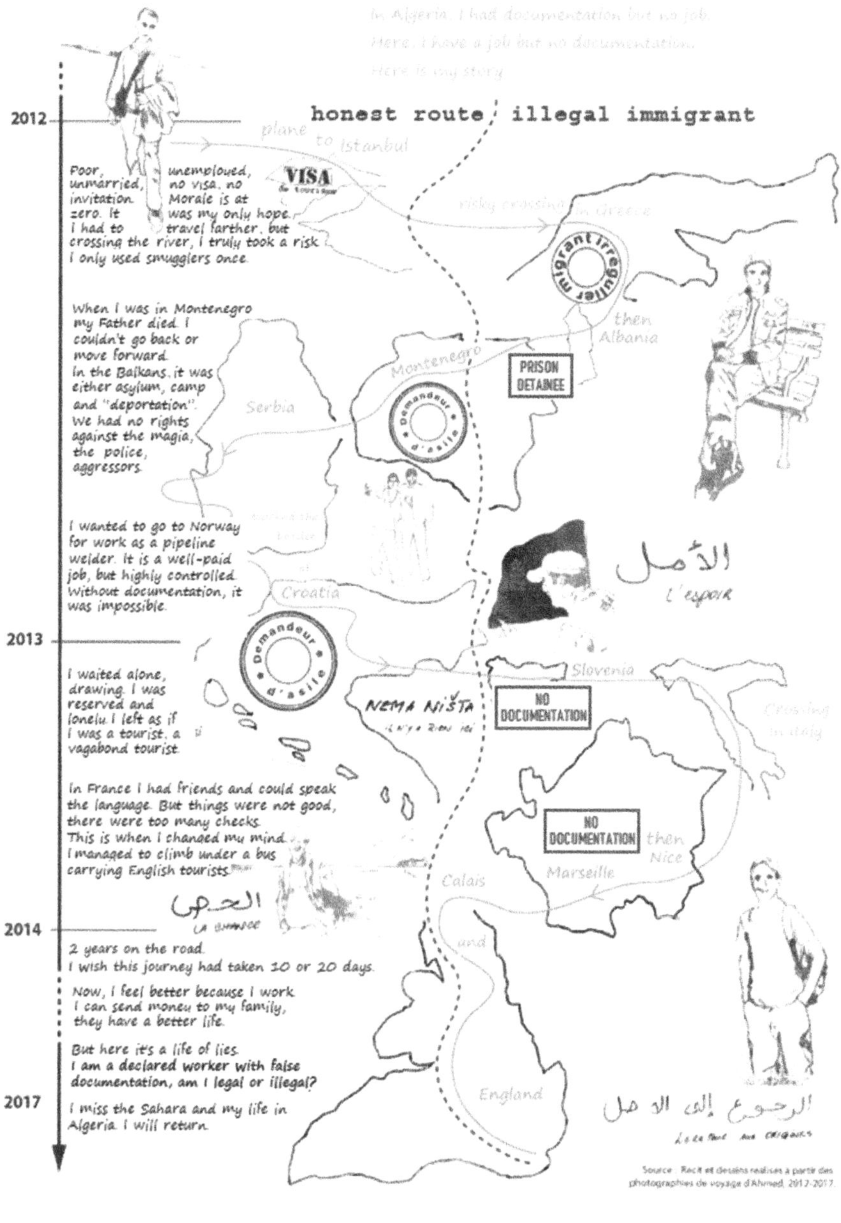

Figure 10.8. *Jobless or undocumented, Ahmed's dilemma (source: map by M. Dujmovic with the assistance of F. Bahoken and O. Clochard for Migreurop in Migreurop (2017, p. 19))*

Nevertheless, one should not overestimate the democratic character of the map in the sense that people do not always understand the intentions of the designers; hence the need to make animations, to tell stories, to bring in the elements that make up the map as it unfolds, to make what I call small cartographic objects. In search of paths toward a double cartographic and political emancipation, the exhibition *Moving Beyond Borders*, conceived with the company *Étrange Miroir*, is part of a militant and artistic approach, in several languages and using a multimedia approach. It conveys "in the plastic and sound work (...) a sensitive experience promoting a vibration, a disposition to be surprised" (Arlais *et al.* 2018). These synergies show how these cartographies elaborated as "kaleidoscopic spatialities" are essential for the emergence of new – and beautiful – ways of conceiving migration and the struggles waged against the policies that oppose it.

10.4. Conclusion

While some maps describe a worldview as strong as philosophy, law or anthropology can, sometimes with irrefutable facts obtained at the cost of significant associative and academic work, we are well aware that government resistance to potential changes remains strong. As Springer *et al.* (2018) point out, "the struggle for freedom and social justice is a difficult undertaking that necessarily involves direct action" (p. 31). Thus, the work and political dimensions of Migreurop's maps are part of advocacy or contentious actions, but they are also concerned with providing knowledge that can come as a counterpoint to that mobilized by institutional, media actors, etc. The idea is to elaborate cartographic proposals, admittedly sometimes incomplete and subjective, but which make it possible to continue to support the two compasses of the network: the one going against the use of camps and the other in favor of freedom of movement, that is to say, "radical political imaginaries capable of demanding the impossible" (Springer *et al.* 2018, p. 14).

10.5. References

Agier, M. (2014). *Un monde de camps*. La Découverte, Paris.

Arlais, M., Bacon, L., Clochard, O., Rialland, R. (2018). Moving beyond borders: la fabrique d'une œuvre scientifique. *Hommes & Migrations*, 1322, 125–134.

Bacon, L., Clochard, O., Honoré, T., Lambert, N., Mekdjian, S., Rekacewicz, P. (2016). Cartographier les mouvements migratoires. *Revue européenne des migrations internationales*, 32(3–4).

Berthomière, W. (2012). En-quête de signes : migrations, places et continuité(s). HDR, University of Poitiers.

Blanchard, E. and Wender, A.-S. (eds) (2007). *Guerre aux migrants. Le livre noir de Ceuta et Melilla*. Éditions Syllepse, Paris.

Boltanski, L. (2007). *La souffrance à distance*. Folio Essais, Paris.

Cauvin, C., Escobar, F., Serradj, A. (2010). *Thematic Cartography and Transformations*. ISTE Ltd, London and John Wiley & Sons, New York.

Charles, C. (2007). Accords de réadmission et respect des droits de l'Homme dans les pays tiers. Bilan et perspective pour le Parlement européen, Étude pour la sous-commission des Droits de l'Homme du Parlement Européen. Brussels [Online]. Available at: http://www.migreurop.org/article1277.html [Accessed 24 November 2014].

Clochard, O. (2018). Mineurs étrangers en rétention administrative : situations indéterminées, *Jeunes et Mineurs en Mobilité*, 3, 31–39 [Online]. Available at: https://omm.hypotheses.org/files/2018/12/JMM3-VL.pdf [Accessed 24 November 2014].

Clochard, O. (2020). Retour sur la réintroduction de la notion controversée de camp dans les sciences sociales. In *Encampés : de quel(s) droit(s)* ? Beulay, M., Chaumette, A.-L., Dubin, L., Eudes, M. (eds). Institut Francophone pour la Justice et la Démocratie, Paris.

European Commission (2006). Rapport de la Commission européenne pour la mise en place d'un nouveau cadre de partenariat avec les pays tiers dans le cadre de l'Agenda européen en matière de migration. Report, European Commission, Brussels.

Gould, P. and Bailly, A. (1995). *Le pouvoir des cartes, Brian Harley et la cartographie*. Economica, Paris.

Lagarde, D. (2019). Sur les routes de l'exil syrien. Récits de vie et parcours migratoires des réfugiés de Deir Mqaren. PhD Thesis, Université Jean Jaurès, Toulouse.

Lambert, N. and Zanin, C. (2019). *Mad Maps – L'atlas qui va changer le monde*. Armand Colin, Paris.

Lévy, J., Poncet, P., Tricoire, E. (eds) (2004). La carte, enjeu contemporain. *La Documentation Photographique*, no. 8036. La Documentation Française, Paris.

Migreurop (2010). Pour la fermeture des camps d'étrangers en Europe et au-delà [Online]. Available at: http://www.migreurop.org/article1718.html [Accessed 24 November 2014].

Migreurop (2012). Accords de réadmission. La "coopération" au service de l'expulsion des migrants. *Les notes de Migreurop*, 1, 1–4 [Online]. Available at: http://www.migreurop.org/IMG/pdf/Note_de_MIGREUROP_12122012_Accords_de_readmission_pour_mise_en_ligne.pdf.

Migreurop (2013). Pour la liberté de circulation [Online]. Available at: http://www.migreurop.org/article2395.html [Accessed 24 November 2014].

Migreurop (2017). *Atlas des migrants en Europe. Approches critiques des politiques migratoires*. Armand Colin, Paris.

Migreurop (2019). L'OIM, une organisation au service des frontières… fermées. *Les notes de Migreurop*, 9, 1–4 [Online]. Available at: http://www.migreurop.org/IMG/pdf/note_9_fr.pdf.

Noiriel, G. (2010). *Dire la vérité au pouvoir. Les intellectuels en question*. Agone, Marseille.

Rekacewicz, P. (2006). La cartographie, entre science, art et manipulation. *Le Monde diplomatique*, 14–15 [Online]. Available at: https://www.monde-diplomatique.fr/2006/02/REKACEWICZ/13169 [Acessed 24 November 2014].

Simon, G. (2006). Migrations, la spatialisation du regard. *Revue européenne des migrations internationales*, 22(2), 9–22 [Online]. Available at: https://doi.org/10.4000/remi.2815 [Accessed 24 November 2014].

Springer, S., Des Rochers, A., Gauthier, A. (2018). *Pour une géographie anarchiste*. Lux Éditeur, Montreal.

Tyszler, E. (2019). Derrière les barrières de Ceuta & Melilla. Rapports sociaux de sexe, de race et colonialité du contrôle migratoire à la frontière maroco-espagnole. PhD Thesis, Paris 8 University Vincennes-Saint-Denis.

List of Authors

Anne BAILLY
Honorary Technical Director,
AEBK, Maps and Communication
and Consultant
France

Federica BURINI
University of Bergamo
Italy

Olivier CLOCHARD
CNRS
Migrinter
University of Poitiers
France

Hy DAO
University of Geneva
Switzerland

Bernard DEBARBIEUX
University of Geneva
Switzerland

Franco FARINELLI
University of Bologna
Italy

Pierre GAUTREAU
UMR PRODIG
University of Paris 1 Panthéon-Sorbonne
France

Irène HIRT
University of Geneva
Switzerland

Sylvie LARDON
INRAE & AgroParisTech
UMR Territoires
Clermont-Ferrand
France

Jacques LÉVY
Director
Spatial Intelligence Chair
Université Polytechnique
Hauts-de-France
France

Matthieu NOUCHER
CNRS
UMR Passages
University of Bordeaux
France

Index

H, I

L, M

N, O, P

R, S

T, V, W

Printed and bound by CPI Group (UK) Ltd, Croydon, CR0 4YY

13/09/2026

14960995-0004